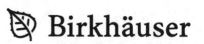

Lev A. Sakhnovich

Integral Equations with Difference Kernels on Finite Intervals

Second Edition, Revised and Extended

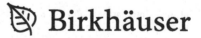 Birkhäuser

Lev A. Sakhnovich
Milford, CT, USA

ISSN 0255-0156   ISSN 2296-4878   (electronic)
Operator Theory: Advances and Applications
ISBN 978-3-319-30763-3   ISBN 978-3-319-16489-2   (eBook)
DOI 10.1007/978-3-319-16489-2

Mathematics Subject Classification (2010): 45A05, 47B38, 47A12, 60G12, 60G52

Springer Cham Heidelberg New York Dordrecht London
First Edition © Springer Basel AG, 1996
© Springer International Publishing Switzerland 2015
Softcover reprint of the hardcover 2nd edition 2015

Printed on acid-free paper

Springer International Publishing AG Switzerland is part of Springer Science+Business Media
(www.birkhauser-science.com)

To the 100*th Anniversary of my Teacher and Friend
Professor V.P. Potapov* (1914–1981)

Contents

Preface to the Second Edition

The following changes are made in the second edition.

1. A number of misprints is corrected.

2. The chapter "Operator Bezoutiant and Roots of Entire Functions" is omitted because the results of this chapter are developed and improved in our book [97, Chap. 11].

3. The text dedicated to Lévy processes is essentially extended and, correspondingly, several new chapters are added (see Chapters 7–10). In particular, the following new important results are included: convolution form of infinitesimal generators corresponding to Lévy processes (Chapter 7), construction of quasi-potential operators in the space of continuous functions and solution of several problems formulated by M. Kac (Chapter 8), results on Cauchy type Lévy processes and new examples of Lévy processes (Chapter 9) as well as results on Lévy processes with summable Lévy measures (Chapter 10).

4. In the last Chapter 11 we formulate some open problems, which are connected with the theory of integral equations.

The author is very grateful to A. Sakhnovich for his help and very useful remarks and to I. Roitberg for her help in the preparation of the manuscript.

Milford, CT, USA,
December 2014

Lev Sakhnovich

Introduction to the First Edition

An equation of the form

$$Sf = \mu f(x) + \int\limits_{\omega_1}^{\omega_2} k(x - t) f(t) \, dt = \varphi(x) \tag{1}$$

where μ is a complex number,

$$k(x) \in L(\omega_1 - \omega_2, \omega_2 - \omega_1), \quad -\infty \le \omega_1 < \omega_2 \le \infty$$

is called an equation with a difference kernel. Equations of this kind are used in numerous applied problems (the problem of optimal synthesis [113], light scattering in the atmosphere [100], diffraction on a ribbon [24], the theory of elasticity [1]). This fact is usually caused by certain homogenity of the considered processes in time or space.

The same equations play an important role in a number of mathematical problems (inverse spectral problems [90], the theory of nonlinear integrable equations [91], and the theory of stable processes [30]).

The equations (1) can be separated into three essentially different classes:

(I) $\omega_1 = -\infty, \quad \omega_2 = \infty$;

(II) $\omega_1 = 0, \quad \omega_2 = \infty$;

(III) $\omega_1 = 0, \quad \omega_2 = \omega < \infty$.

The solution of equation (1) of the class (I) can be found with the help of the Fourier transform [104]. N. Wiener and E. Hopf [26] as well as V.A. Fock [14] introduced an effective factorization method for solving equation (1) of class (II). This method was later perfected and completed by M.G. Kreĭn in his paper [40].

The specifics of various applied problems (the finite memory of an apparatus, the finite domain of a contact) leads to the necessity of investigating equation (1) on the finite segment (case (III)). Our book is dedicated to this important and difficult case in which equation (1) takes the form

$$Sf = \mu f(x) + \int\limits_{0}^{\omega} k(x - t) f(t) \, dt = \varphi(x). \tag{2}$$

In Kreĭn's paper [38], an equation of this type was solved by factorizing the operator S. In the works on astrophysics (Ambartsumian [2], Sobolev [100], Chandrasekhar [13], Ivanov [29]), we come across the detailed study of the case

$$\varphi(x) = e^{i\lambda x}. \tag{3}$$

All these works exploit the close connection of the operator S with the operator of differentiation:

$$\left(S\frac{d}{dx}f - \frac{d}{dx}Sf \right) = 0, \tag{4}$$

where the function $f(x)$ is supposed to be continuously differentiable and $f(0) = f(\omega) = 0$.

In this book we consider the class of equations

$$Sf = \frac{d}{dx}\int_0^\omega s(x-t)f(t)\,dt = \varphi(x). \tag{5}$$

Formula (5) gives the general form of the bounded operator with a difference kernel. Setting in (5)

$$s(x) = \int_0^x k(u)\,du + \mu_+ \quad (x > 0),$$

$$s(x) = \int_0^x k(u)\,du + \mu_- \quad (x < 0),$$

$$\mu = \mu_+ - \mu_-,$$

we obtain operators of the class (2).

The representation of the operator S in the form (5) allows us to consider uniformly different classes of the operators S and of the corresponding operator equations $Sf = \varphi$. For the confirmation of this fact we give a list of the main special cases of formula (5) and also indicate the chapters where these special cases are considered:

(I) $$Sf = \sum_{j=N}^M \mu_j f(x - x_j) + \int_0^\omega K(x-t)f(t)\,dt,$$

where $f(x) = 0$ for $x \notin [0, \omega]$, $K(x) \in L(-\omega, \omega)$ (see Chapter 1);

(II) $$Sf = \mu f(x) + \int_0^\omega K(x-t)f(t)\,dt,$$

where $K(x) \in L(-\omega, \omega)$, $\mu \neq 0$ (see Chapters 1, 2, and 3);

(III)
$$Sf = \int_0^\omega K(x - t) f(t) \, dt,$$

where $K(x) \in L(-\omega, \omega)$ (see Chapters 1–4);

(IV)
$$Sf = \int_0^\omega \left(\frac{1}{x-t} + K(x-t) \right) f(t) \, dt,$$

where $K(x) \in L(-\omega, \omega)$ (see Chapter 3);

(V)
$$Sf = \frac{1}{\Gamma(i\alpha + 1)} \frac{d}{dx} \int_0^x (x - t)^{i\alpha} f(t) \, dt,$$

(see Chapters 1 and 3). Thus, we apply a unified method to different equations with a difference kernel both of the first and of the second kind. Besides the transition from the equation (2) to the general form (5), we also use the integral operator A:

$$Af = i \int_0^x f(t) \, dt, \quad A^* f = -i \int_x^\omega f(t) \, dt \tag{6}$$

instead of the differential operator in (4). The analogue of the formula (4) is the operator identity

$$(AS - SA^*) f = i \int_0^\omega (M(x) + N(t)) f(t) \, dt, \tag{7}$$

where $M(x) = s(x)$, $N(x) = -s(-x)$, $0 \leq x \leq \omega$.

The formula (7) is a particular case of the operator identity of the form

$$AS - SB = Q. \tag{8}$$

In the present book our method of operator identities is used for solving the equation (5), that is, for the construction of the operator $T = S^{-1}$. Assuming that (8) holds, we apply the following scheme in order to construct T. From (8) we have

$$TA - BT = TQT. \tag{9}$$

Hence, we can recover T from (9) when the right-hand side of (9) is known. We apply this approach to (7). Then the operator $T = S^{-1}$ can be explicitly recovered from the functions $N_1(x)$ and $N_2(x)$, which are defined by the relations

$$SN_1 = M, \quad SN_2 = 1. \tag{10}$$

The obtained results permit us to clarify the structure of the operators $T = S^{-1}$ (see Chapters 1 and 2) and in many cases to construct the operators $T = S^{-1}$ explicitly (see Chapter 3). The knowledge of N_1 and N_2 is the minimal information, which is necessary in order to construct $T = S^{-1}$.

An important role in the theory of the equations with a difference kernel is played by the equations with a special right-hand side

$$S\big(B(x,\lambda)\big) = e^{i\alpha x}. \tag{11}$$

We note that the reflection coefficient in the light scattering theory coincides with the function

$$\rho(\lambda,\mu) = \int_0^\omega B(x,\lambda)e^{i\mu x}\,\mathrm{d}x. \tag{12}$$

Here we describe the analytical structure of $\rho(\lambda,\mu)$ and deduce the following generalization of the well-known Ambartsumian formula [2]:

$$\rho(\lambda,\mu) = -ie^{i\omega\mu}\frac{a(\lambda)b(-\mu) - b(\lambda)a(-\mu)}{\lambda + \mu}, \tag{13}$$

where

$$a(\lambda) = i\lambda \int_0^\omega e^{i\lambda t} N_2(\omega - t)\,\mathrm{d}t,$$

$$b(\lambda) = e^{i\lambda\omega} - i\lambda \int_0^\omega e^{i\lambda t} N_1(\omega - t)\,\mathrm{d}t.$$

We clarify the connection of the function $\rho(\lambda,\mu)$ with the theory of transfer matrix functions. General results are followed by the analysis of a number of important examples (Keldysh–Lavrentjev equation [34]; Galin-Shmatkova equation, see, e.g., [11]; and others). The classical equations

$$\int_0^\omega \ln \frac{A}{2|x - t|} f(t)\,\mathrm{d}t = \varphi(x), \tag{14}$$

$$\int_0^\omega \frac{1 - \beta\,\mathrm{sgn}\,(x \doteq t)}{|x - t|^h} f(t)\,\mathrm{d}t = \varphi(x) \tag{15}$$

are studied in detail.

We remark that the operator T is an operator analogue of the Bezoutiant. This fact allows us to deduce a number of theorems about the distribution of the roots of entire functions which develop further the results by M.G. Kreĭn [42] (see Chapter 4).

The problems of the contact theory of elasticity [102] and of the diffraction theory [24] lead to equations of the form

$$S_\Delta f = \mu f(x) + \int_\Delta k(x-t)f(t)\,\mathrm{d}t = \varphi(x), \quad x \in \Delta,$$

where Δ denotes the system of segments. The operator S_Δ can be transformed into an operator with a W-difference kernel. The operator with a W-difference kernel is a certain generalization of the operator with a difference kernel. More precisely, it is an operator of the form

$$Sf = \frac{\mathrm{d}}{\mathrm{d}x} \int_0^\omega s(x,t)f(t)\,\mathrm{d}t, \tag{16}$$

where $f(t) = \mathrm{col}\begin{bmatrix} f_1 & f_2 & \cdots & f_n \end{bmatrix}$ and the matrix kernel $s(x,t)$ has the form

$$s(x,t) = \left\{ s_{l,m}(\omega_l x - \omega_m t) \right\}_{l,m=1}^n. \tag{17}$$

If $\omega_1 = \omega_2 = \cdots = \omega_n$, then the operator S is an operator with a matrix difference kernel. For the operator S with a W-difference kernel, the operator identity of the form (7) is fulfilled. On the basis of this operator identity, the procedure of construction of the operator $T = S^{-1}$ is given (see Chapter 5).

We apply equations with a difference kernel to the theory of stochastic stable processes. The homogeneous process $X(\tau)$ $(X(0) = 0)$ with independent increments is called a stable process if

$$E\left[\exp\left(i\xi X(\tau)\right)\right] = \exp\left(-\tau|\xi|^\alpha \left(1 - i\beta(\,\mathrm{sgn}\,\xi)\tan\frac{\pi\alpha}{2}\right)\right),$$

where E is the mathematical expectation and $0 < \alpha \le 2$, $-1 \le \beta \le 1$. Stable processes are a natural generalization of Wiener processes ($\alpha = 2$).

It was shown by M. Kac and H. Pollard [33] that the study of these processes leads to the solution of integro-differential equations. However, as M. Kac wrote in [30], the solution of the obtained equations presents great analytical difficulties. M. Kac solved this equation for the Cauchy symmetrical process. In this book we investigated all the symmetrical stable processes and obtained answers to all the problems by M. Kac [30, §8] concerning stable processes. Some results are extended to the case of non-symmetric stable processes.

An essential place in the book belongs to the application of general theoretical results to concrete problems of physics and technics. We give a physical interpretation of a number of results, and obtain new formulas in the theories of elasticity, hydrodynamics and communication.

In some chapters we formulated open problems in order to outline the limits of the existing theory and to stimulate an independent research of the readers.

I should like to express my gratitude to S. Pozin and A. Sakhnovich for their help in the preparation of the manuscript. I am also thankful to my wife E. Melnichenko for translating the book into English.

I am grateful to Prof. I. Gohberg for prompting me to write this book and for his careful editing.

The work on this book was partially supported by the International Science Foundation-Ukraine (G. Soros Foundation).

Chapter 1

Invertible Operator with a Difference Kernel

In this chapter, we deal with the problem of inversion of the operators S acting in the space $L^2(0, \omega)$ and having the form

$$Sf = \frac{\mathrm{d}}{\mathrm{d}x} \int_0^\omega s(x - t) f(t) \, \mathrm{d}t, \quad f(x) \in L^2(0, \omega) \tag{1.0.1}$$

where $s(x) \in L^2(-\omega, \omega)$ and the function

$$g(x) = \int_0^\omega s(x - t) f(t) \, \mathrm{d}t$$

is assumed to be absolutely continuous.

An operator S that can be represented in the form (1.0.1) is called an operator with a difference kernel. In particular, the following operator can be reduced to the form (1.0.1):

$$Sf = \sum_{j=N}^M \mu_j f(x - x_j) + \int_0^\omega k(x - t) f(t) \, \mathrm{d}t, \tag{1.0.2}$$

where $f(x) = 0$ for $x \notin [0, \omega]$, $x_j \in (-\omega, \omega)$. In this case $s(x)$ is a function of a bounded variation without a singular component, and

$$s'(x) = k(x) \in L(-\omega, \omega), \quad s(x_j + 0) - s(x_j - 0) = \mu_j.$$

When we put $\mu_j = 0$ for $j \neq 0$ and $x_0 = 0$, we have

$$Sf = \mu_0 f(x) + \int_0^\omega k(x - t) f(t) \, \mathrm{d}t, \quad k(x) \in L(-\omega, \omega). \tag{1.0.3}$$

Many operators with difference kernels that arise in concrete problems have the form (1.0.3). However, operators with difference kernels that are not representable in the form (1.0.3) are also used in a number of questions (generalized stationary processes [19], the distribution of the roots of entire functions). Consider, for instance, the operator

$$Sf = \mathcal{J}^{i\alpha} f = \frac{1}{\Gamma(i\alpha + 1)} \frac{\mathrm{d}}{\mathrm{d}x} \int_0^x (x - t)^{i\alpha} f(t)\,\mathrm{d}t, \quad 0 \le x \le \omega, \qquad (1.0.4)$$

where $\alpha = \overline{\alpha}$ and $\Gamma(z)$ is a gamma function.

By analogy with the fractional integration [25], the operator $\mathcal{J}^{i\alpha}$ is called an integration operator of order $i\alpha$ [35, 70]. The function $s(x)$ in case of (1.0.4) has the form

$$s(x) = \begin{cases} 0, & x < 0, \\ x^{i\alpha}/\Gamma(i\alpha + 1), & x > 0. \end{cases}$$

Since $s'(x) \notin L(-\omega, \omega)$, $\mathcal{J}^{i\alpha}$ cannot be presented in the form (1.0.3).

The transfer from the form (1.0.3) to (1.0.1) does not entail additional difficulties. The more general form (1.0.1) outlines the natural boundaries of the theory and leads to more complete results.

Our approach to the inversion of operators with a difference kernel was published first in [76, 77] (see also [64, 78, 84, 89] and references therein). Further on, an important role is played by the functions $N_1(x)$, $N_2(x)$ which satisfy the relations

$$SN_1(x) = M(x), \quad SN_2(x) = \mathbb{1}, \qquad (1.0.5)$$

where $\mathbb{1}$ stands for the function which identically equals 1 and

$$M(x) = s(x), \qquad 0 \le x \le \omega. \qquad (1.0.6)$$

The operator $T = S^{-1}$ is expressed explicitly via $N_1(x)$ and $N_2(x)$. Thus, in order to construct the operator T, it is sufficient to know how it acts on $\mathbb{1}$ and $M(x)$. These results make it possible to describe the structure of the class of operators $T = S^{-1}$.

The action of the operator T on the function $e^{i\lambda x}$ is considered in greater detail. That is, we study the equation

$$Sf = e^{i\lambda x}. \qquad (1.0.7)$$

Here we obtain analogies with the well-known Hopf–Fock formulas [14, 26] for a semi-infinite interval. The corresponding facts for a finite interval and an operator (1.0.3) were first obtained in the works on astrophysics [2, 13, 29, 100].

1.1 Constructing the inverse operator

1. We need the following theorem.

Theorem 1.1.1. *Any bounded operator S acting in $L^2(0,\omega)$ can be represented in the form*

$$Sf = \frac{d}{dx} \int_0^\omega s(x,t) f(t) \, dt \tag{1.1.1}$$

where $s(x,t)$ belongs to $L^2(0,\omega)$ for each fixed x.

Proof. We introduce the function

$$e_x = \begin{cases} 1, & t \le x, \\ 0, & t > x. \end{cases} \tag{1.1.2}$$

If $f \in L^2(0,\omega)$, then $Sf \in L^2(0,\omega)$ whenever S is bounded. It is immediate that

$$\langle Sf, e_x \rangle = \int_0^x (Sf) \, dt,$$

where $\langle \cdot, \cdot \rangle$ denotes the scalar product in $L^2(0,\omega)$, that is,

$$\langle f, g \rangle = \int_0^\omega \overline{g(x)} f(x) \, dx. \tag{1.1.3}$$

Besides, the equality

$$\langle Sf, e_x \rangle = \langle f, S^* e_x \rangle \tag{1.1.4}$$

is true where the operator S^* is an adjoint one in relation to the operator S. Setting

$$\overline{S^* e_x} = s(x,t), \tag{1.1.5}$$

we have

$$\langle f, S^* e_x \rangle = \int_0^\omega s(x,t) f(t) \, dt. \tag{1.1.6}$$

It follows from (1.1.4)–(1.1.6) that

$$\int_0^x (Sf) \, dt = \int_0^\omega s(x,t) f(t) \, dt. \tag{1.1.7}$$

By differentiating both parts of (1.1.7) we obtain the desired representation. $\square$

It follows from Theorem 1.1.1 that formula (1.0.1) defines the most general form of a bounded operator with a difference kernel. For unitary operators the representation (1.1.1) was obtained by Bochner (see [62, Chap. 7, §3]).

From the definition of e_x and equality (1.1.5) we deduce the following corollary.

Corollary 1.1.2. *The function $s(x,t)$ in (1.1.1) can be chosen so that $s(x,t)$ belongs to $L^2(0,\omega)$ for each x and*

$$s(0,t) = 0, \qquad \int_0^\omega |s(x + \Delta x, t) - s(x,t)|^2 \, dt \le \|S\|^2 |\Delta x|.$$

2. We consider the integral operator

$$Af = i \int_0^x f(t) \, dt. \tag{1.1.8}$$

Then the adjoint operator A^* has the form

$$A^* f = -i \int_x^\omega f(t) \, dt. \tag{1.1.9}$$

The next result [84] plays an essential role in what follows.

Theorem 1.1.3. *For any bounded operator S with a difference kernel, we have the operator identity*

$$(AS - SA^*) f = i \int_0^\omega \Big(M(x) + N(t) \Big) f(t) \, dt, \tag{1.1.10}$$

where

$$M(x) = s(x), \quad N(x) = -s(-x), \qquad 0 \le x \le \omega. \tag{1.1.11}$$

Proof. From (1.0.1) and (1.1.8), (1.1.9) it follows that

$$(AS - SA^*) f = i \int_0^\omega s(x - t) f(t) \, dt - i \int_0^\omega s(-t) f(t) \, dt$$

$$+ i \frac{d}{dx} \int_0^\omega s(x - u) \int_u^\omega f(t) \, dt \, du. \tag{1.1.12}$$

We calculate the double integral on the right-hand side of (1.1.12)

$$\frac{d}{dx} \int_0^\omega s(x - u) \int_u^\omega f(t) \, dt \, du = \int_0^\omega \Big(s(x) - s(x - t) \Big) f(t) \, dt.$$

Substituting this expression in (1.1.12) and using the notation (1.1.11) we obtain the theorem. □

Remark 1.1.4. A result on operators S satisfying identities of the form (1.1.10), which is an inverse statement to the statement of Theorem 1.1.3, is given in Theorem 1.2.3.

Remark 1.1.5. The operator identities of the form (1.1.10) permit us to investigate a large amount of analysis problems (see [84, 90, 91, 93, 94, 97]).

3. Suppose, that the operator S of the form (1.0.1) has the bounded inverse operator $T = S^{-1}$. In this book the method of operator identities of the form (1.1.10) is the basis for investigating and constructing the operator $T = S^{-1}$. As

$$T(AS - SA^*)T = TA - A^*T,$$

then from the operator identity (1.1.10), we deduce that

$$(TA - A^*T)f = i \int_0^\omega \left(N_1(x)\overline{M_1(t)} + N_2(x)\overline{M_2(t)} \right) f(t)\,dt, \qquad (1.1.13)$$

where

$$S^*M_1 = \mathbb{1}, \quad S^*M_2 = \overline{N(x)}, \quad SN_1 = M(x), \quad SN_2 = \mathbb{1}. \qquad (1.1.14)$$

Let $N_k(x)$, $M_k(x)$ $(k = 1, 2)$ be a function in $L^2(0, \omega)$. We introduce the function

$$Q(x, t) = N_1(x)\overline{M_1(t)} + N_2(x)\overline{M_2(t)} \qquad (1.1.15)$$

and the operator

$$Qf = \int_0^\omega Q(x, t)f(t)\,dt. \qquad (1.1.16)$$

The following Theorem 1.1.6 is valid for all operators Q of the form (1.1.16) such that the functions $Q(x, t)$ satisfy natural summability conditions. In particular, it is valid for Q such that $Q(x, t) = g_1(x)g_2(t)$, where g_1 and g_2 are row and column vectors, respectively, and their entries belong $L^2(0, \omega)$. (Thus, the result holds for the case that $Q(x, t)$ is given by (1.1.15).)

Theorem 1.1.6 ([84]). *If a bounded operator T acting in $L^2(0, \omega)$ satisfies the operator equation*

$$TA - A^*T = iQ, \qquad (1.1.17)$$

then the function

$$\Phi(x, t) = \frac{1}{2} \int_{x+t}^{2\omega - |x-t|} Q\left(\frac{s + x - t}{2}, \frac{s - x + t}{2} \right) ds \qquad (1.1.18)$$

is absolutely continuous in t and

$$Tf = \frac{\mathrm{d}}{\mathrm{d}x} \int_0^{\omega} \left(\frac{\partial}{\partial t} \Phi(x,t) \right) f(t) \, \mathrm{d}t. \tag{1.1.19}$$

Proof. Since T is bounded, Theorem 1.1.1 implies that

$$Tf = \frac{\mathrm{d}}{\mathrm{d}x} \int_0^{\omega} F(x,t) f(t) \, \mathrm{d}t. \tag{1.1.20}$$

According to Corollary 1.1.2 the function $F(x,t)$ can be chosen so that

$$F(\omega, t) = 0, \quad \int_0^{\omega} |F(x + \Delta x, t) - F(x,t)|^2 \, \mathrm{d}t \leq \|T\|^2 |\Delta x|. \tag{1.1.21}$$

Thus, the integral $\int_t^{\omega} F(x,s) \, \mathrm{d}s$ is continuous in x, that is, the iterated integral

$$F_1(x,t) = - \int_x^{\omega} \int_t^{\omega} F(u,s) \, \mathrm{d}s \, \mathrm{d}u \tag{1.1.22}$$

is well defined. It follows from (1.1.20), (1.1.22) that the operator

$$T_1 f = \int_0^{\omega} F_1(x,t) f(t) \, \mathrm{d}t \tag{1.1.23}$$

satisfies the relation

$$T_1 = \mathrm{i} A^{*2} T A. \tag{1.1.24}$$

Taking into account (1.1.17), we deduce the equality

$$T_1 A - A^* T_1 = \mathrm{i} Q_1, \tag{1.1.25}$$

where

$$Q_1 f = \mathrm{i} A^{*2} Q A f = \int_0^{\omega} Q_1(x,t) f(t) \, \mathrm{d}t, \tag{1.1.26}$$

$$Q_1(x,t) = \int_x^{\omega} \int_t^{\omega} Q(u,s) \, \mathrm{d}s \big((u-x) \, \mathrm{d}u \big). \tag{1.1.27}$$

The operator equality (1.1.25) means that

$$\int\limits_{t}^{\omega} F_1(x,s)\,ds + \int\limits_{x}^{\omega} F_1(s,t)\,ds = Q_1(x,t). \tag{1.1.28}$$

From (1.1.22), (1.1.28) there follows the equality

$$\int\limits_{t}^{\omega} \frac{\partial F_1(x,s)}{\partial x}\,ds - F_1(x,t) = \frac{\partial Q_1}{\partial x}. \tag{1.1.29}$$

By (1.1.27) and (1.1.29) we have

$$-\frac{\partial F_1(x,t)}{\partial x} - \frac{\partial F_1(x,t)}{\partial t} = \frac{\partial^2 Q_1(x,t)}{\partial t \partial x}. \tag{1.1.30}$$

Using (1.1.22) we obtain the relation

$$F_1(\omega,t) = F_1(x,\omega) = 0. \tag{1.1.31}$$

We introduce the notations

$$F_2(\xi,\eta) = F_1\left(\frac{\xi+\eta}{2}, \frac{\xi-\eta}{2}\right), \quad Q_2(x,t) = \frac{\partial^2 Q_1(x,t)}{\partial x \partial t} = \int\limits_{x}^{\omega} Q(u,t)\,du. \tag{1.1.32}$$

Having substituted variables $\xi = x + t$, $\eta = x - t$ in (1.1.30), (1.1.31) we obtain

$$-2\frac{\partial F_2(\xi,\eta)}{\partial \xi} = Q_2\left(\frac{\xi+\eta}{2}, \frac{\xi-\eta}{2}\right), \quad F_2(2\omega - |\eta|, \eta) = 0 \tag{1.1.33}$$

from which the equality

$$F_2(\xi,\eta) = -\frac{1}{2}\int\limits_{2\omega - |\eta|}^{\xi} Q_2\left(\frac{s+\eta}{2}, \frac{s-\eta}{2}\right)\,ds \tag{1.1.34}$$

directly follows.

Coming back in (1.1.34) to the variables x and t, we have

$$F_1(x,t) = -\frac{1}{2}\int\limits_{2\omega - |x-t|}^{x+t} Q_2\left(\frac{s+x-t}{2}, \frac{s-x+t}{2}\right)\,ds. \tag{1.1.35}$$

We introduce the variable $z = (s - x + t)/2$ and rewrite (1.1.35) in the form

$$F_1(x,t) = \begin{cases} -\int\limits_{\omega - x + t}^{t} Q_2(z + x - t, z)\,dz, & x \geq t, \\[3mm] \int\limits_{t}^{\omega} Q_2(z + x - t, z)\,dz, & x \leq t. \end{cases}$$

Then, taking into account (1.1.32) we obtain

$$\frac{\partial F_1}{\partial x} = \int_{\omega-x+t}^{t} Q(z+x-t,z)\,\mathrm{d}z, \quad x \geq t,$$

$$\frac{\partial F_1}{\partial x} = -\int_{t}^{\omega} Q(z+x-t,z)\,\mathrm{d}z, \quad x \leq t.$$

Thus, the equality

$$\frac{\partial F_1}{\partial x} = -\frac{1}{2} \int_{x+t}^{2\omega-|x-t|} Q\left(\frac{s+x-t}{2}, \frac{s-x+t}{2}\right)\,\mathrm{d}s \qquad (1.1.36)$$

is proved.

By (1.1.22)

$$F(x,t) = -\frac{\partial^2 F_1(x,t)}{\partial t \partial x}. \qquad (1.1.37)$$

Combining (1.1.20), (1.1.36), (1.1.37) we obtain the theorem. $\qquad\square$

Introduce the involution operator

$$Uf = \overline{f(\omega - x)}. \qquad (1.1.38)$$

We easily check that

$$UAU = A^*, \qquad (1.1.39)$$

which is also a particular case of Lemma 1.1.9. Thus, if S satisfies the operator identity

$$AS - SA^* = \mathrm{i}Q, \qquad (1.1.40)$$

then $T = USU$ satisfies the identity $TA - A^*T = \mathrm{i}UQU$. Therefore we may apply Theorem 1.1.6 in order to solve (1.1.40).

Corollary 1.1.7. *If a bounded operator S acting in $L^2(0,\omega)$ satisfies the operator equation (1.1.40), then the function*

$$\Phi(x,t) = \frac{1}{2} \int_{|x-t|}^{x+t} Q\left(\frac{s+x-t}{2}, \frac{s-x+t}{2}\right)\,\mathrm{d}s \qquad (1.1.41)$$

is absolutely continuous in t and

$$Sf = \frac{\mathrm{d}}{\mathrm{d}x} \int_{0}^{\omega} \left(\frac{\partial}{\partial t}\Phi(x,t)\right) f(t)\,\mathrm{d}t. \qquad (1.1.42)$$

Corollary 1.1.7 is used in the study of several classes of structured operators (and not only operators with difference kernels), which are essential in various inverse problems (see, e.g., [16, 65–69]).

Remark 1.1.8. Let the operator S with a difference kernel have a bounded inverse $T = S^{-1}$. Then according to (1.1.13) the operator T satisfies the relation (1.1.17). It means that the operator T can be constructed explicitly using (1.1.15), (1.1.18) and (1.1.19). Here we suppose that the functions $N_1(x)$, $N_2(x)$, $M_1(x)$, $M_2(x)$ defined by the relations (1.1.14) are known.

4. There exist simple connections between the functions $N_1(x)$, $N_2(x)$, $M_1(x)$, $M_2(x)$. In order to show it we need the lemma below.

Lemma 1.1.9. *The relation*

$$USU = S^* \tag{1.1.43}$$

is true.

Proof. Let $g(x)$ be a differentiable function such that $g(0) = g(\omega) = 0$. Then

$$\langle Sf, g \rangle = \int_0^\omega \frac{d}{dx} \int_0^\omega s(x-t)f(t)\,dt\overline{g(x)}\,dx$$

$$= -\int_0^\omega \int_0^\omega s(x-t)f(t)\,dt\overline{g'(x)}\,dx$$

$$= -\int_0^\omega f(t) \int_0^\omega s(x-t)\overline{g'(x)}\,dx\,dt,$$

hence,

$$S^*g = -\int_0^\omega \overline{s(t-x)}g'(x)\,dt = -\frac{d}{dx}\int_{-x}^{\omega-x} \overline{s(v)}g(v+t)\,dv. \tag{1.1.44}$$

Setting $v + x = t$, we rewrite (1.1.44) in the form

$$S^*g = -\frac{d}{dx}\int_0^\omega \overline{s(t-x)}g(t)\,dt. \tag{1.1.45}$$

It is easy to see that

$$USUg = -\frac{d}{dx}\int_0^\omega \overline{s(t-x)}g(t)\,dt. \tag{1.1.46}$$

Consequently, (1.1.43) is satisfied on the set functions g, which is dense in $L^2(0, \omega)$. Since S is bounded, the relation holds for all g in $L^2(0, \omega)$. □

Lemma 1.1.10. *Suppose that there exist N_1 and N_2 in $L^2(0, \omega)$ such that*

$$SN_1 = M, \quad SN_2 = \mathbb{1}. \tag{1.1.47}$$

Then

$$S^* M_1 = \mathbb{1}, \quad S^* M_2 = \overline{N(x)} \tag{1.1.48}$$

where

$$M_1(x) = \overline{N_2(\omega - x)}, \quad M_2(x) = \mathbb{1} - \overline{N_1(\omega - x)}. \tag{1.1.49}$$

Proof. A direct calculation shows that

$$S\mathbb{1} = M(x) + N(\omega - x) \tag{1.1.50}$$

that is

$$S\big(\mathbb{1} - N_1(x)\big) = N(\omega - x). \tag{1.1.51}$$

Now (1.1.48) and (1.1.49) follow immediately from (1.1.47), (1.1.51) and (1.1.43).

$\square$

From (1.1.15) and Lemma 1.1.10 we have

Theorem 1.1.11. *Let the operator S of the form* (1.0.1) *act in $L^2(0, \omega)$ and have a bounded inverse operator. Then the operator T is defined by* (1.1.18), (1.1.19) *where*

$$Q(x, t) = N_2(\omega - t)N_1(x) + \big(\mathbb{1} - N_1(\omega - t)\big)N_2(x). \tag{1.1.52}$$

Thus, for constructing the operators $T = S^{-1}$, it is sufficient to know how it acts on the functions $\mathbb{1}$ and $M(x)$.

5. Suppose, in addition, that S with a difference kernel is a self-adjoint operator. Then by (1.1.45)

$$\frac{\mathrm{d}}{\mathrm{d}x} \int_0^\omega \left(s(x - t) + \overline{s(t - x)} \right) f(t) \, \mathrm{d}t = 0, \quad f \in L^2(0, \omega),$$

that is

$$\int_0^\omega \left(s(x - t) + \overline{s(t - x)} \right) f(t) \, \mathrm{d}t = C_f.$$

According to the Riesz theorem [62] on linear functionals, there exists such a function $\mu(t) \in L^2(0, \omega)$, that

$$C_f = \int_0^\omega \mu(t) f(t) \, \mathrm{d}t.$$

It follows from the last two equalities that

$$\mu(t) = s(x - t) + \overline{s(t - x)},$$

that is,

$$\mu(t) = \text{const} = \mu = \bar{\mu}.$$

Thus, for the self-adjoint operator S, the equality

$$s(x) + \overline{s(-x)} = \mu \tag{1.1.53}$$

is true. It follows from our reasoning that for the operator S to be self-adjoint the condition (1.1.52) is not only necessary but sufficient as well. Using the notation (1.1.21) we rewrite the condition (1.1.53) in the form

$$M(x) - \overline{N(x)} = \mu, \quad \mu = \bar{\mu}. \tag{1.1.54}$$

Since the kernel $s(x)$ of S is determined to a constant, we may assume, without loss of generality, that $\mu = 0$. Thus, the condition for S to be self-adjoint has the form

$$M(x) = \overline{N(x)}. \tag{1.1.55}$$

From Lemma 1.1.10 and (1.1.55) it follows that for the self-adjoint operator S the equalities

$$\overline{N_2(\omega - x)} = N_2(x), \quad \mathbb{1} - \overline{N_1(\omega - x)} = N_1(x) \tag{1.1.56}$$

are true.

6. We discuss separately triangular operators with difference kernels:

$$Sf = \frac{d}{dx} \int_0^x s(x - t)f(t)\, dt, \quad f \in L^2(0, \omega). \tag{1.1.57}$$

Theorem 1.1.12. *If S and its inverse are bounded, and if S has the form* (1.1.57), *then*

$$S^{-1}f = \frac{d}{dx} \int_0^x N_2(x - t)f(t)\, dt, \quad f \in L^2(0, \omega) \tag{1.1.58}$$

where $N_2(x) = S^{-1}\mathbb{1}$.

Proof. A direct calculation shows that $N_1(x) = \mathbb{1}$. Then by (1.1.52) we have $Q(x, t) = N_2(\omega - t)$. By (1.1.18) the relations

$$\frac{\partial}{\partial t}\Phi(x, t) = \begin{cases} N_2(x - t) - N_2(\omega - t), & \text{for} \quad x > t, \\ -N_2(\omega - t), & \text{for} \quad x < t \end{cases}$$

are true. Now (1.1.58) follows from (1.1.19). $\qquad \square$

1.2 Existence conditions and the structure of the inverse operator

1. Let us find conditions under which S is invertible. We denote the range of S by R_S.

Lemma 1.2.1. *Suppose that* $\mathbb{1}$ *and* $M(x)$ *belong to* R_S. *Then* R_S *is dense in* $L^2(0, \omega)$.

Proof. We verify that there is a sequence of functions $\mathcal{L}_m(x)$ for which

$$S\mathcal{L}_m = x^{m-1}, \quad m = 1, 2, \ldots \tag{1.2.1}$$

If in (1.1.10) we put $f = \mathcal{L}_m$, then it follows from (1.1.8) and (1.2.1) that

$$\mathrm{i}\frac{x^m}{m} = SA^*\mathcal{L}_m + \mathrm{i}\int_0^\omega \Big(M(x) + N(t)\Big)\mathcal{L}_m(t)\,\mathrm{d}t. \tag{1.2.2}$$

Since $\mathbb{1}$, $M \in R_S$ there exist N_1 and N_2 such that the relations (1.1.47) are true. We can, therefore, write

$$\mathrm{i}\frac{x^m}{m} = S\left(A^*\mathcal{L}_m + \mathrm{i}\int_0^\omega \Big(N_1(x) + N(t)N_2(x)\Big)\mathcal{L}_m(t)\,\mathrm{d}t\right)$$

that is,

$$\frac{1}{m}\mathcal{L}_{m+1} = -\int_x^\omega \mathcal{L}_m(t)\,\mathrm{d}t + \int_0^\omega \Big(N_1(x) + N(t)N_2(x)\Big)\mathcal{L}_m(t)\,\mathrm{d}t. \tag{1.2.3}$$

According to the conditions of the lemma, there is a function $\mathcal{L}_1(x) = N_2(x)$. The relation (1.2.3) determines all the other terms of the sequence $\mathcal{L}_m$, $m = 1, 2, \ldots$. Thus, $x^m \in R_S$ for $m = 0, \ldots$, and the lemma follows. $\square$

Theorem 1.2.2. *Suppose that* $\mathbb{1}$ *and* $M(x)$ *belong to* R_S *and the operator* T *determined by* (1.1.18), (1.1.19) *and* (1.1.52) *is bounded. Then* S *is invertible and* $T = S^{-1}$.

Proof. We rewrite (1.2.3) in the form

$$\frac{1}{m}\mathcal{L}_{m+1} = \int_0^\omega t^{m-1}N_2(\omega - t)\,\mathrm{d}t\, N_1(x)$$

$$+ \int_0^\omega t^{m-1}\Big(\mathbb{1} - N_1(\omega - t)\Big)\,\mathrm{d}t N_2(x) - \int_x^\omega \mathcal{L}_m(t)\,\mathrm{d}t. \tag{1.2.4}$$

Here we have used the equalities

$$\int_0^\omega t^{m-1} N_2(\omega - t) \, \mathrm{d}t = \left\langle S\mathcal{L}_m, \overline{N_2(\omega - t)} \right\rangle = \int_0^\omega \mathcal{L}_m(t) \, \mathrm{d}t, \qquad (1.2.5)$$

$$\int_0^\omega t^{m-1} \left(\mathbb{1} - N_1(\omega - t)\right) \mathrm{d}t = \left\langle S\mathcal{L}_m, M_2 \right\rangle = \int_0^\omega \mathcal{L}_m(t) N(t) \, \mathrm{d}t. \qquad (1.2.6)$$

Next we introduce the sequence of functions

$$L_m(x) = T x^{m-1}; \quad m = 1, 2, \ldots \qquad (1.2.7)$$

It follows from (1.1.18), (1.1.19) that

$$(TA - A^*T)f = \mathrm{i} \int_0^\omega Q(x, t) f(t) \, \mathrm{d}t. \qquad (1.2.8)$$

If we put $f(x) = x^{m-1}$, in (1.2.8) then it follows from (1.1.52) and (1.2.7), that

$$\frac{1}{m} L_{m+1}(x) = \int_0^\omega t^{m-1} N_2(\omega - t) \, \mathrm{d}t N_1(x)$$

$$+ \int_0^\omega t^{m-1} \left(\mathbb{1} - N_1(\omega - t)\right) \mathrm{d}t N_2(x) - \int_x^\omega L_m(t) \, \mathrm{d}t. \qquad (1.2.9)$$

Using (1.1.18), (1.1.19) we calculate now

$$T\mathbb{1} = -\frac{1}{2} \frac{\mathrm{d}}{\mathrm{d}x} \int_x^{2\omega - x} Q\left(\frac{s + x}{2}, \frac{s - x}{2}\right) \mathrm{d}s.$$

Taking into account (1.1.52), we obtain

$$L_1 = T\mathbb{1} = -\frac{1}{2} \frac{\mathrm{d}}{\mathrm{d}x} \int_x^{2\omega - x} N_2\left(\frac{s + x}{2}\right) \mathrm{d}s = -\frac{\mathrm{d}}{\mathrm{d}x} \int_x^\omega N_2(v) \, \mathrm{d}v = N_2(x). \qquad (1.2.10)$$

It follows from (1.2.4), (1.2.9) and (1.2.10) that $L_m = \mathcal{L}_m$. Then the equality $STx^{m-1} = x^{m-1}$ is true, that is,

$$ST = I. \qquad (1.2.11)$$

Multiplying both parts of (1.2.11) on the left and on the right by U and taking into account (1.1.43), we obtain $S^*UTU = I$, that is,

$$UT^*US = I. \qquad (1.2.12)$$

It follows from (1.2.11), (1.2.12) that the operator S is invertible, $T = S^{-1}$ and

$$T = UT^*U. \tag{1.2.13}$$

Thus, the theorem is proved.   □

2. We need an inverse one to Theorem 1.1.3.

Theorem 1.2.3. *If a bounded operator S acting in $L^2(0, \omega)$ satisfies*

$$\left(AS - SA^*\right)f = \mathrm{i} \int_0^\omega \Big(M(x) + N(t)\Big) f(t)\, \mathrm{d}t \tag{1.2.14}$$

for some $M(x)$ and $N(x)$ in $L^2(0, \omega)$, then S is an operator with a difference kernel and (1.1.11) holds.

Proof. The relation (1.2.14) can be written in the following form:

$$\left(S_1 A - A^* S_1\right)f = \mathrm{i} \int_0^\omega \Big(\overline{M(\omega - x)} + \overline{N(\omega - t)}\Big) f(t)\, \mathrm{d}t \tag{1.2.15}$$

where $S_1 = USU$, $A^* = UAU$. Applying Theorem 1.1.6 to the operator equation (1.2.12) we see that

$$S_1 f = \frac{\mathrm{d}}{\mathrm{d}x} \int_0^\omega \left(\frac{\partial}{\partial t}\Phi_1(x, t)\right) f(t)\, \mathrm{d}t. \tag{1.2.16}$$

Here

$$\Phi_1(x, t) = \frac{1}{2} \int_{x+t}^{2\omega - |x-t|} \left(\overline{M\left(\omega - \frac{s+x-t}{2}\right)} + \overline{N\left(\omega - \frac{s-x+t}{2}\right)}\right)\, \mathrm{d}s,$$

that is

$$\Phi_1(x, t) = \begin{cases} \int_x^\omega \overline{M(\omega - s)}\, \mathrm{d}s + \int_t^{\omega - x + t} \overline{N(\omega - s)}\, \mathrm{d}s, & x > t, \\ \int_x^{\omega + x - t} \overline{M(\omega - s)}\, \mathrm{d}s + \int_t^\omega \overline{N(\omega - s)}\, \mathrm{d}s, & x < t. \end{cases} \tag{1.2.17}$$

Putting (1.2.17) on the right-hand side of (1.2.16) we have

$$S_1 f = \frac{\mathrm{d}}{\mathrm{d}x} \int_0^\omega s_1(x - t) f(t)\, \mathrm{d}t,$$

where
$$s_1(x) = \overline{N(x)}, \quad -s_1(-x) = M(x), \quad 0 \le x \le \omega. \tag{1.2.18}$$
Using $US_1U = S$ we derive the final formula

$$Sf = \frac{d}{dx} \int_0^\omega s(x-t)f(t)\,dt, \quad s(x) = -s_1(-x). \tag{1.2.19}$$

The statement of the theorem follows from (1.2.18), (1.2.19). $\qquad$ □

3. Let us clarify the structure of the class of operators $T = S^{-1}$, where S is a bounded operator with a difference kernel.

Theorem 1.2.4. *Suppose that both an operator T acting in $L^2(0,\omega)$ and its inverse are bounded. Then $S = T^{-1}$ is an operator with a difference kernel if and only if T is determined by* (1.1.18), (1.1.19) *and* (1.1.52) *where $N_1, N_2 \in L^2(0,\omega)$.*

Proof. The *necessity* of the conditions follows from the preceding arguments (see Section 1.1).

Sufficiency. To calculate T we consider

$$\langle Tf, \mathbb{1} \rangle = \int_0^\omega f(t)\frac{d}{dt}\left(\overline{\Phi(\omega,t) - \Phi(0,t)}\right) dt,$$

that is,

$$T^*\mathbb{1} = \frac{d}{dt}\left(\overline{\Phi(\omega,t) - \Phi(0,t)}\right) = -\frac{1}{2}\frac{d}{dt}\int_t^{2\omega-t} Q\left(\frac{s-t}{2}, \frac{s+t}{2}\right) ds.$$

Using (1.1.52) we obtain
$$T^*\mathbb{1} = \overline{N_2(\omega - x)}. \tag{1.2.20}$$

Now for $S = T^{-1}$ by (1.1.52) and (1.2.8), we have

$$(AS - SA^*)f = i\left\langle Sf, U(N_2)SN_1 + \left\langle Sf, U(\mathbb{1} - N_1)\right\rangle SN_2 \right\rangle.$$

Hence, in accordance with (1.2.10) and (1.2.20) we deduce

$$\left(AS - SA^*\right)f = i\int_0^\omega \left(M(x) + N(t)\right)f(t)\,dt, \tag{1.2.21}$$

where $M(x) = SN_1$, $\overline{N(x)} = S^*\left(\mathbb{1} - \overline{N_1(\omega - x)}\right)$.

According to Theorem 1.2.3 a solution of (1.2.21) has the form (1.0.1), where $s(x)$ is determined by (1.1.11). This proves the theorem. $\qquad$ □

1.3   Equations with a special right-hand side

In Section 1.1 we constructed the operator $T = S^{-1}$ from the functions $N_1(x)$ and $N_2(x)$. Using these results we consider the equation

$$Sf = e^{i\lambda x}. \tag{1.3.1}$$

Theorem 1.3.1. *Let S be a bounded operator with a difference kernel and suppose that there are functions $N_1(x), N_2(x) \in L^2(0, \omega)$ such that*

$$SN_2 = \mathbb{1}, \quad SN_1 = M(x). \tag{1.3.2}$$

Then

$$SB(x, \lambda) = e^{ix\lambda}, \tag{1.3.3}$$

where

$$B(x, \lambda) = u(x, \lambda) - i\lambda \int_x^\omega e^{i(x-t)\lambda} u(t, \lambda)\,dt, \tag{1.3.4}$$

$$u(x, \lambda) = a(\lambda)N_1(x) + b(\lambda)N_2(x), \tag{1.3.5}$$

$$a(\lambda) = i\lambda \int_0^\omega e^{i\lambda t} N_2(\omega - t)\,dt, \quad b(\lambda) = e^{i\lambda \omega} - i\lambda \int_0^\omega e^{i\lambda t} N_1(\omega - t)\,dt. \tag{1.3.6}$$

Proof. From (1.2.3) we have for a certain C

$$\|\mathcal{L}_{m+1}\| \leq Cm\|\mathcal{L}_m\|,$$

that is,

$$\|\mathcal{L}_{m+1}\| \leq C^{m+1} m!. \tag{1.3.7}$$

We put

$$B(x, \lambda) = \sum_{m=0}^{\infty} \frac{(i\lambda)^m}{m!} \mathcal{L}_{m+1}. \tag{1.3.8}$$

By (1.3.7), the series on the right-hand side of (1.3.8) converges for $|\lambda| < C^{-1}$. Since S is bounded,

$$SB(x, \lambda) = \sum_{m=0}^{\infty} \frac{(i\lambda)^m}{m!} x^m = e^{ix\lambda}, \quad |\lambda| < C^{-1}. \tag{1.3.9}$$

From (1.2.3) and (1.3.8) it follows that

$$B(x, \lambda) = u(x, \lambda) - i\lambda \int_x^\omega B(t, \lambda)\,dt, \tag{1.3.10}$$

where $u(x, \lambda)$ is determined by (1.3.5), (1.3.6). When we solve this Volterra equation, we obtain (1.3.4) for $|\lambda| < C^{-1}$. Since $B(x, \lambda)$ and $e^{i x \lambda}$ are analytic in λ and S is bounded, it follows that (1.3.9) holds for all λ. This proves the theorem. $\qquad\square$

Results close to Theorem 1.3.1 for operators S of the form (1.0.3) were first obtained in connection with astrophysical problems [29, 100].

2. An essential role in the theory of operators with difference kernels is played by the function

$$\rho(\lambda, \mu) = \int_0^\omega e^{i \mu x} B(x, \lambda)\, dx. \qquad (1.3.11)$$

We derive a simple representation for $\rho(\lambda, \mu)$. Using (1.3.4), we have that

$$\rho(\lambda, \mu) = \int_0^\omega e^{i t \mu} u(t, \lambda)\, dt - i\lambda \int_0^\omega \int_x^\omega e^{i(x-t)\lambda} u(z, \lambda)\, dt e^{i x \mu}\, dx,$$

that is,

$$\rho(\lambda, \mu) = \int_0^\omega u(t, \lambda) \frac{\mu e^{i t \mu} + \lambda e^{-i t \lambda}}{\lambda + \mu}\, dt.$$

Taking into account (1.3.5), we obtain

$$\rho(\lambda, \mu) = \int_0^\omega \Big(a(\lambda) N_1(\omega - t) + b(\lambda) N_2(\omega - t) \Big) \frac{\mu e^{i(\omega - t)\mu} + \lambda e^{-i(\omega - t)\lambda}}{\lambda + \mu}\, dt.$$

Hence, by (1.3.6) we have the final formula

$$\rho(\lambda, \mu) = -i e^{i \omega \mu} \frac{a(\lambda) b(-\mu) - b(\lambda) a(-\mu)}{\lambda + \mu}. \qquad (1.3.12)$$

Let us assume in addition that $N_1(x)$ and $N_2(x)$ are continuous at $x = 0$ and $x = \omega$. Then by (1.3.4), the Ambartsumian functions [2]

$$X(\lambda) = B(0, \lambda), \quad y(\lambda) = B(\omega, \lambda) \qquad (1.3.13)$$

are well defined. It follows from (1.3.4) and (1.3.13) that

$$X(\lambda) = a(\lambda)\big(N_1(0) - \mathbb{1}\big) + b(\lambda) N_2(0), \qquad (1.3.14)$$
$$Y(\lambda) = a(\lambda) N_1(\omega) + b(\lambda) N_2(\omega). \qquad (1.3.15)$$

Now (1.3.14) and (1.3.15) are analogous to the Hopf–Fock formulas [14, 26] for a semi-infinite interval. We write

$$v = \big(-N_1(0) + \mathbb{1}\big) N_2(\omega) + N_2(0) N_1(\omega). \qquad (1.3.16)$$

Assuming that $v \neq 0$ we deduce from (1.3.14) and (1.3.15) that

$$a(\lambda) = -\frac{1}{v}\Big(X(\lambda)N_2(\omega) - Y(\lambda)N_2(0)\Big), \tag{1.3.17}$$

$$b(\lambda) = \frac{1}{v}\Big(X(\lambda)N_1(\omega) - Y(\lambda)\big(N_1(0) - 1\big)\Big). \tag{1.3.18}$$

Substituting (1.3.17) and (1.3.18) in (1.3.12), we obtain

$$\rho(\lambda, \mu) = \frac{ie^{i\omega\mu}}{v(\lambda + \mu)} \Big(X(\lambda)Y(-\mu) - Y(\lambda)X(-\mu)\Big) \tag{1.3.19}$$

which plays an essential role in the theory of radiation transfer.

    Other derivations of this formula for operators of the form (1.0.3) are contained in the works by Ambartsumian [2], Sobolev [100] and Ivanov [29]. Formula (1.3.12) clarifies the analytic structure of the function $\rho(\lambda, \mu)$. This function of two variables can be expressed by the functions of one variable $a(\lambda)$ and $b(\lambda)$.

1.4   Operators with difference kernel in $L^p(0, \omega)$

1. By analogy with Theorem 1.1.1 we can prove the following proposition:

Any bounded operator S acting in $L^p(0, \omega)$ $(p \geq 1)$ can be represented in the form

$$Sf = \frac{d}{dx} \int_0^\omega s(x, t)f(t)\, dt,$$

where $s(x, t)$ for every fixed x belongs to $L^q(0, \omega)$ $\left(\dfrac{1}{p} + \dfrac{1}{q} = 1\right)$. From this it follows that

$$Sf = \frac{d}{dx} \int_0^\omega s(x - t)f(t)\, dt, \quad s(x) \in L^q(-\omega, \omega) \tag{1.4.1}$$

defines the most general form of a bounded operator in $L^p(0, \omega)$ $(p \geq 1)$ with a difference kernel.

2. In what follows we use $\|S\|_p$ to denote the norm of the operator S in the space $L^p(0, \omega)$. If $f \in L^p(0, \omega)$ and $g \in L^q(0, \omega)$ their scalar product is defined by

$$\langle f, g \rangle = \int_0^\omega f(x)\overline{g(x)}\, dx. \tag{1.4.2}$$

Theorem 1.4.1. *Suppose that an operator S defined by (1.4.1) is bounded in $L^p(0, \omega)$ $(1 \leq p \leq 2)$. Then S is bounded in all the spaces $L^r(0, \omega)$ $(p \leq r \leq q)$ and*

$$\|S\|_p = \|S\|_q, \quad \|S\|_r \leq \|S\|_p, \quad p \leq r \leq q. \tag{1.4.3}$$

Proof. Let $f \in L^p(0, \omega)$, $g \in L^q(0, \omega)$. The adjoint of the operator S is denoted by S^* and is defined by the equality

$$\langle Sf, g \rangle = \langle f, S^*g \rangle. \tag{1.4.4}$$

Hence,

$$\|S\|_p = \|S^*\|_q. \tag{1.4.5}$$

Repeating the arguments of Lemma 1.1.9, we obtain

$$S^* = USU, \tag{1.4.6}$$

that is,

$$\|S\|_q = \|S^*\|_q. \tag{1.4.7}$$

The first relation (1.4.3) follows from (1.4.5) and (1.4.7). Then the second relation (1.4.3) follows from the Rietz–Thorin interpolation theorem [106]. $\quad\square$

As an example we consider an operator of the form

$$Sf = \int_0^\omega k(x - t)f(t)\, \mathrm{d}t, \quad k(x) \in L(-\omega, \omega). \tag{1.4.8}$$

It is easy to check that

$$\|S\|_1 \leq \int_{-\omega}^\omega |k(x)|\, \mathrm{d}x$$

from which we conclude, by (1.4.3), that

$$\|S\|_r \leq \int_{-\omega}^\omega |k(x)|\, \mathrm{d}x. \tag{1.4.9}$$

We note that it is easy to deduce (1.4.9) directly [40].

3. We consider separately the case when $N_1(x)$ and $N_2(x)$ are absolutely continuous.

Theorem 1.4.2. *Suppose that the functions $N_1(x)$ and $N_2(x)$ are absolutely continuous. Then the operator T determined by (1.1.18), (1.1.19), (1.1.52) admits the representation*

$$Tf = vf + \int_0^\omega \gamma(x,t) f(t) \, dt \tag{1.4.10}$$

and there exists a function $h(x)$ in $L(-\omega, \omega)$ such that

$$|\gamma(x,t)| \le h(x-t), \quad 0 \le x, t \le \omega. \tag{1.4.11}$$

Proof. We write

$$A(x) = \begin{bmatrix} N_2(x) & 1 - N_1(x) \end{bmatrix}, \quad B(x) = \begin{bmatrix} N_1(x) \\ N_2(x) \end{bmatrix}. \tag{1.4.12}$$

By (1.1.18) and (1.1.52) we have

$$\frac{\partial \Phi}{\partial t}\bigg|_{x=t+0} - \frac{\partial \Phi}{\partial t}\bigg|_{x=t-0} = A(0)B(\omega), \tag{1.4.13}$$

$$\frac{\partial^2 \Phi}{\partial x \partial t} = A'(x-t)B(\omega) - \int_t^{\omega+t-x} A'(\omega - s)B'(s + x - t) \, ds \quad (x > t), \tag{1.4.14}$$

$$\frac{\partial^2 \Phi}{\partial x \partial t} = -A(0)B'(\omega + x - t) - \int_t^\omega A'(\omega - s)B'(s + x - t) \, ds \quad (x < t). \tag{1.4.15}$$

Putting

$$v = A(0)B(\omega), \quad \gamma(x,t) = \frac{\partial^2 \Phi}{\partial x \partial t} \tag{1.4.16}$$

we obtain (1.4.10) from (1.1.19) and (1.4.12)–(1.4.15). Now (1.4.11) is satisfied if

$$h(x) = \begin{cases} \left| A'(x)B(\omega) \right| + \int_0^{\omega-x} \left| A'(\omega - s)B'(s + x) \right| ds, & 0 < x < \omega, \\ \left| A(0)B'(\omega + x) \right| + \int_{-x}^{\omega} \left| (\omega - s)B'(s + x) \right| ds, & -\omega < x < 0. \end{cases}$$

From (1.4.9) and (1.4.11) it follows that

$$\|T\|_p \le |v| + \int_{-\omega}^{\omega} |h(x)| \, dx. \tag{1.4.17}$$

$\square$

4. We now return to the operator S.

Theorem 1.4.3. *Suppose that an operator S with a difference kernel is bounded in $L(0, \omega)$ and that there exist absolutely continuous functions $N_1(x)$ and $N_2(x)$ such that*

$$SN_1 = M(x), \quad SN_2 = 1. \tag{1.4.18}$$

Then:

I. *S and its inverse are bounded in $L^p(0, \omega)$ for all $p \geq 1$.*

II. *S admits a representation*

$$Sf = \mu f(x) + \int_0^\omega k(x - t)f(t)\, dt, \tag{1.4.19}$$

where

$$\mu \neq 0, \quad k(x) \in L(-\omega, \omega). \tag{1.4.20}$$

III. *The operator $S^{-1} = T$ is determined by (1.4.10) and (1.4.16).*

Proof. By Theorem 1.4.1, S is bounded on all the spaces $L^p(0, \omega)$, $p \geq 1$. Then, by Theorem 1.2.2, T is invertible in $L^2(0, \omega)$. Since by (1.4.10), (1.4.14)–(1.4.16) the integral term of T is completely continuous, we have

$$v = N_2(0)N_1(\omega) + \left(1 - N_1(0)\right)N_2(\omega) \neq 0. \tag{1.4.21}$$

A direct calculation gives

$$SN_k = \frac{d}{dx} \int_{x-\omega}^x N_k(x - u)s(u)\, du$$

$$= N_k(0)s(x) - N_k(\omega)s(x - \omega) + \int_0^\omega N_k'(t)s(x - t)\, dt; \quad k = 1, 2. \tag{1.4.22}$$

Taking (1.4.18) and (1.4.22) into account we obtain

$$\int_0^\omega N_1'(t)s(x - t)\, dt = \left(1 - N_1(0)\right)s(x) + N_1(\omega)s(x - \omega), \tag{1.4.23}$$

$$\int_0^\omega N_2'(t)s(x - t)\, dt = 1 - N_2(0)s(x) + N_2(\omega)s(x - \omega). \tag{1.4.24}$$

Since S is bounded in $L(0,\omega)$ and by virtue of (1.4.23), (1.4.24) we have

$$\frac{\mathrm{d}}{\mathrm{d}x}\left(\left(\mathbb{1} - N_1(0)\right)s(x) + N_1(\omega)s(x - \omega)\right) \in L(0,\omega),$$

$$\frac{\mathrm{d}}{\mathrm{d}x}\left(\mathbb{1} - N_2(0)\,s(x) + N_2(\omega)\,s(x - \omega)\right) \in L(0,\omega),$$

that is, the functions

$$\left(\mathbb{1} - N_1(0)\right)s(x) + N_1(\omega)\,s(x - \omega)$$

and

$$\mathbb{1} - N_2(0)\,s(x) + N_2(\omega)\,s(x - \omega)$$

are absolutely continuous on $[0,\omega]$. Hence, using (1.4.21) we deduce that

$$s'(x), s'(x - \omega) \in L(0,\omega).$$

Thus, (1.4.19) holds, where

$$\mu = s(+0) - s(-0) = \frac{1}{v}, \quad k(x) \in s'(x) \tag{1.4.25}$$

that is, statements I, II of this theorem are true. Now statement III follows from Theorem 1.4.2. □

5. In this subsection we derive a result of Gohberg and Sementsul [22].

Theorem 1.4.4. *Let S be an operator of the form*

$$Sf = f(x) + \int_0^\omega k(x - t)f(t)\,\mathrm{d}t, \quad k(x) \in L(-\omega, \omega).$$

If there are functions $\gamma_\pm(x) \in L(0,\omega)$ for which

$$S\gamma_+ = k(x), \quad S\gamma_- = k(x - \omega) \tag{1.4.26}$$

then S is invertible in $L^p(0,\omega)$, $p \geq 1$ and its inverse $T = S^{-1}$ has the form

$$Tf = f(x) + \int_0^\omega \gamma(x,t)f(t)\,\mathrm{d}t, \tag{1.4.27}$$

where

$$\gamma(x,t) = -\gamma_+(x-t) \tag{1.4.28}$$

$$- \int_t^{\omega+t-x} \Big(\gamma_-(\omega-s)\gamma_+(s+x-t) - \gamma_+(\omega-s)\gamma_-(s+x-t)\Big)\,ds,$$

$$x > t,$$

$$\gamma(x,t) = -\gamma_-(\omega+x-t) \tag{1.4.29}$$

$$- \int_t^{\omega} \Big(\gamma_-(\omega-s)\gamma_+(s+x-t) - \gamma_+(\omega-s)\gamma_-(s+x-t)\Big)\,ds,$$

$$x < t.$$

Proof. Let us start with the case $k(x) = 0$ for $0 \le x \le \omega$. Then the operators S and T have the form

$$Sf = f(x) + \int_x^\omega k(x-t)f(t)\,dt,$$

$$Tf = f(x) - \int_x^\omega \gamma_-(\omega+x-t)f(t)\,dt.$$

By (1.4.26) we have

$$\gamma_-(x) + \int_x^\omega k(x-t)\gamma_-(t)\,dt = k(x-\omega).$$

This equality can be written for $x < t$ in the following form

$$\gamma_-(\omega+x-t) + \int_x^t k(x-u)\gamma_-(\omega+u-t)\,du = k(x-t).$$

This means that the operators S and T are reciprocally inverse and the theorem in this case is true. We also consider that

$$k(x) \neq 0, \quad 0 \le x \le \omega. \tag{1.4.30}$$

We write L_S for the subspace of $L(0,\omega)$ mapped to zero by S. Then $\dim L_S = n < \infty$. We write $\varphi_1, \varphi_2, \ldots, \varphi_n$ for a linearly independent system in L_S. By (1.1.10) we can write

$$S\left(\int_x^\omega \varphi_k(t)\,dt \right) = \int_0^\omega \varphi_k(t)\Big(M(x) + N(t)\Big)\,dt. \tag{1.4.31}$$

Further, from (1.1.10), (1.1.50) and (1.4.26) it follows that

$$
S\left(\int_x^\omega \gamma_+(t)\,dt\right) = \alpha_1 M(x) + \beta_1, \quad S\left(\int_x^\omega \gamma_-(t)\,dt - \mathbb{1}\right) = \alpha_2 M(x) + \beta_2,
$$

$$(1.4.32)$$

where

$$
\alpha_1 = \int_0^\omega \gamma_+(t)\,dt - 1, \quad \alpha_2 = \int_0^\omega \gamma_-(t)\,dt - 1. \tag{1.4.33}
$$

Let us verify that the functions

$$
\int_x^\omega \varphi_m(t)\,dt \quad (1 \le m \le n), \qquad \int_x^\omega \gamma_+(t)\,dt, \qquad \int_x^\omega \gamma_-(t)\,dt - \mathbb{1} \tag{1.4.34}
$$

are linearly independent. Indeed, from

$$
\sum_{m=1}^n C_m \int_x^\omega \varphi_m(t)\,dt + C_{n+1} \int_x^\omega \gamma_+(t)\,dt + C_{n+2}\left(\int_x^\omega \gamma_-(t)\,dt - \mathbb{1}\right) = 0 \quad (1.4.35)
$$

it follows for $x = \omega$ that $C_{n+2} = 0$. Differentiating (1.4.35) and applying S, we see that $C_{n+1}k(x) = 0$, that is, $C_{n+1} = 0$. Now the linear independence of $\varphi_1, \varphi_2, \ldots, \varphi_n$ implies that $C_1 = C_2 = \cdots = C_n = 0$. Thus, the system (1.4.34) is linearly independent. Since $\dim L_S = n$, the operator S takes this system to a subspace of dimension not less than two. On the other hand, it follows from (1.4.31) and (1.4.32) that this subspace is spanned by $M(x)$ and $\mathbb{1}$. Consequently,

$$
\Delta = \alpha_1\beta_2 - \alpha_2\beta_1 \ne 0. \tag{1.4.36}
$$

From (1.4.32) and (1.4.36) it follows that

$$
N_1(x) = \frac{1}{\Delta}\left(\beta_2 \int_x^\omega \gamma_+(t)\,dt - \beta_1\left(\int_x^\omega \gamma_-(t)\,dt - \mathbb{1}\right)\right), \tag{1.4.37}
$$

$$
N_2(x) = \frac{1}{\Delta}\left(-\alpha_2 \int_x^\omega \gamma_+(t)\,dt + \alpha_1\left(\int_x^\omega \gamma_-(t)\,dt - \mathbb{1}\right)\right). \tag{1.4.38}
$$

Thus, the functions N_1 and N_2 are absolutely continuous. By Theorem 1.4.3 the operator S is invertible in $L^p(0,\omega)$ $(p \ge 1)$. Writing (1.4.12), we deduce from (1.4.37) and (1.4.38) that

$$
A'(x) = \frac{1}{\Delta}\left[\gamma_+(x) \quad \gamma_-(x)\right]\mathcal{L}_1, \quad B'(x) = \frac{1}{\Delta}\mathcal{L}_2\begin{bmatrix}\gamma_+(x)\\\gamma_-(x)\end{bmatrix}, \tag{1.4.39}
$$

where the matrices $\mathcal{L}_1$ and $\mathcal{L}_2$ are defined by the equalities

$$\mathcal{L}_1 = \begin{bmatrix} \alpha_2 & \beta_2 \\ -\alpha_1 & -\beta_1 \end{bmatrix}, \quad \mathcal{L}_2 = \begin{bmatrix} -\beta_2 & \beta_1 \\ \alpha_2 & -\alpha_1 \end{bmatrix}. \tag{1.4.40}$$

By virtue of (1.4.16), we have $A(0)B(\omega) = 1$. On the other hand, according to (1.4.33), (1.4.37) and (1.4.38), we have $A(0)B(\omega) = 1/\Delta$. Hence, $\Delta = 1$. By direct calculation we obtain

$$\mathcal{L}_1 \mathcal{L}_2 = \begin{bmatrix} 0 & -1 \\ 1 & 0 \end{bmatrix}, \quad \mathcal{L}_1 B(\omega) = \begin{bmatrix} -1 \\ 0 \end{bmatrix}, \quad A(0)\mathcal{L}_2 = \begin{bmatrix} 0 & 1 \end{bmatrix}.$$

The equalities (1.4.27)–(1.4.29) now follow from (1.4.10) and (1.4.14)–(1.4.16). This proves the theorem. $\qquad\square$

1.5 The use of the Fourier transform

1. The Fourier transform plays an essential role [40, 104] when studying operators with a difference kernel on $(-\infty, \infty)$ and $(0, \infty)$. This transformation is also useful in the case of a finite segment.

Theorem 1.5.1. *Let the operator S of the form* (1.0.1) *be bounded in the space $L^2(0, \omega)$ and there exist in $L^2(0, \omega)$ the functions $N_1(x)$, $N_2(x)$ satisfying* (1.1.47). *If the functions $a(\lambda)$ and $b(\lambda)$ determined by* (1.3.6) *are bounded on the axis $-\infty < \lambda < \infty$, then the operator S has in $L^2(0, \omega)$ a bounded inverse and*

$$\|S^{-1}\| \leq \sup_{-\infty < \lambda < \infty} \left(|a(\lambda)|^2 + |b(\lambda)|^2 \right).$$

Proof. We introduce the operator P_ω determined on the functions of the form

$$\mathcal{H}(\mu) = \int\limits_{-\infty}^{\infty} h(t)\, e^{i\mu t}\, dt, \quad h(t) \in L^2(-\infty, \infty)$$

by means of the formula

$$P_\omega \mathcal{H}(\mu) = \int\limits_{0}^{\omega} h(t)\, e^{i\mu t}\, dt.$$

We denote by H_ω the subspace into which the space $L^2(-\infty, \infty)$ is mapped by the operator P_ω. We introduce the matrices

$$G_+(\lambda) = \begin{bmatrix} a(\lambda) & -b(\lambda) \end{bmatrix}, \quad G_-(\lambda) = -\begin{bmatrix} b(\lambda) & a(\lambda) \end{bmatrix} e^{-i\omega\lambda}.$$

Then (1.3.12) obtains the form

$$\rho(\lambda, \mu) = i \frac{G_-(-\mu)\, G_+^\tau(\lambda)}{\lambda + \mu},$$
(1.5.1)

where G^τ is a matrix transposed in relation to G. Putting

$$\Phi_0(\lambda, \mu) = \int_0^\omega e^{i(\lambda+\mu)t}\, dt = \frac{i}{\lambda + \mu}\left(1 - e^{i(\lambda+\mu)\omega}\right).$$
(1.5.2)

We write

$$G_+^\tau(-\mu)\Phi_0(\lambda, \mu) = \frac{i}{\lambda + \mu}\left(G_+^\tau(-\mu) - G_+^\tau(\lambda)\right)\left(1 - e^{i(\lambda+\mu)\omega}\right)$$
$$+ \frac{i}{\lambda + \mu}G_+^\tau(\lambda)\left(1 - e^{i(\lambda+\mu)\omega}\right).$$

We apply the operator P_ω to both sides of the last equation

$$P_\omega\left(G_+^\tau(-\mu)\Phi_0(\lambda, \mu)\right) = -\frac{i}{\lambda + \mu}\left(G_+^\tau(-\mu) - G_+^\tau(\lambda)\right)e^{i(\lambda+\mu)\omega}$$
$$+ \frac{i}{\lambda + \mu}G_+^\tau(\lambda)\left(1 - e^{i(\lambda+\mu)\omega}\right)$$
(1.5.3)
$$= -\frac{i}{\lambda + \mu}\left(G_+^\tau(-\mu)e^{i(\lambda+\mu)\omega} - G_+^\tau(\lambda)\right).$$

Since $G_-(-\mu)G_+^\tau(-\mu) = 0$, it follows from (1.5.1) and (1.5.3) that

$$\rho(\lambda, \mu) = G_-(-\mu)P_\omega\left(G_+^\tau(-\mu)\Phi_0(\lambda, \mu)\right).$$
(1.5.4)

We define the operator on the functions $F(\mu)$ from $L^2(-\infty, \infty)$:

$$\widetilde{T}F = G_-(-\mu)P_\omega\left(G_+^\tau(-\mu)F(\mu)\right).$$
(1.5.5)

Since $\|P_\omega\| = 1$ and formula (1.5.5) holds, we have the following estimate

$$\|\widetilde{T}\| \le \sup_{-\infty < \lambda < \infty}\left(\left|a(\lambda)\right|^2 + \left|b(\lambda)\right|^2\right).$$
(1.5.6)

Now (1.5.4) implies that

$$\rho(\lambda, \mu) = \widetilde{T}\Phi_0(\lambda, \mu).$$
(1.5.7)

By virtue of (1.3.11) and (1.5.2) the functions $\rho(\lambda, \mu)$ and $\Phi_0(\lambda, \mu)$ belong to H_ω. Besides the linear combinations $\Phi_0(\lambda, \mu)$ are dense in H_ω. Then from (1.5.7) we conclude that the operator $\widetilde{T}$ maps H_ω into itself.

Let us define the operator T using the equality

$$T = V^{-1}\widetilde{T}V, \tag{1.5.8}$$

where V is the Fourier transform

$$Vf = \frac{1}{\sqrt{2\pi}} \int_{-\infty}^{\infty} f(t)e^{i\mu t}\, dt. \tag{1.5.9}$$

According to (1.5.6), (1.5.8), the operator T is bounded. It maps the set of functions, which are equal to zero outside of the segment $[0, \omega]$, into itself. We assume that the operator T acts in $L^2(0, \omega)$. By virtue of (1.5.7) we have

$$Te^{i\lambda x} = B(x, \lambda). \tag{1.5.10}$$

It means that the relation

$$ST = I \tag{1.5.11}$$

holds. Hence, we have the equality $ST = I$. Using (1.1.43) we rewrite the last relation in the form

$$UT^*US = I. \tag{1.5.12}$$

From (1.5.11) and (1.5.12) it follows that the operator S is invertible and that

$$T = S^{-1}. \tag{1.5.13}$$

From the estimate (1.5.6) and the equalities (1.5.8), (1.5.9) and (1.5.13) we deduce the assertion of the theorem. $\qquad\square$

Let us remark that $a(\lambda)$ and $b(\lambda)$ are bounded on the axis $(-\infty, \infty)$ if the functions $N_1(x)$ and $N_2(x)$ have bounded variations.

2. Now (1.5.5) gives the representation of the operator $T = S^{-1}$ in the space of Fourier images. We write a corresponding representation for the operator S defined by the relation (1.0.1). Let us write

$$\widetilde{S}(\mu) = \int_{-\omega}^{\omega} s(t)e^{i\mu t}\, dt. \tag{1.5.14}$$

From

$$Sf = \varphi, \tag{1.5.15}$$

using the properties of the Fourier transform, we obtain

$$\Phi(\mu) = P_\omega\Big(-i\mu F(\mu)\widetilde{S}(\mu)\Big), \tag{1.5.16}$$

where

$$\Phi(\mu) = \int_0^\omega \varphi(t) e^{i\mu t} \, dt, \quad F(\mu) = \int_0^\omega f(t) e^{i\mu t} \, dt. \tag{1.5.17}$$

Thus, the operator S in the space of Fourier images has the form

$$\widetilde{S} F(\mu) = P_\omega \left(-i\mu F(\mu) \widetilde{S}(\mu) \right). \tag{1.5.18}$$

Theorem 1.5.2. *For the boundedness of the operator S of the form* (1.0.1) *in the space $L^2(0,\omega)$ it is sufficient that the function $\lambda \widetilde{S}(\lambda)$ on the axis $-\infty < \lambda < \infty$ is bounded and it is necessary that the function $\lambda \widetilde{S}_1(\lambda)$, where*

$$\widetilde{S}_1(\lambda) = \int_{-\omega}^\omega s(t) \left(\mathbb{1} - \frac{|t|}{\omega} \right) e^{i\lambda t} \, dt, \tag{1.5.19}$$

is bounded on the axis $-\infty < \lambda < \infty$.

Proof. Sufficiency. Since $\|P_\omega\| = 1$, from (1.5.18) it follows that

$$\|S\| \leq \sup \left| \lambda \widetilde{S}(\lambda) \right|, \quad -\infty < \lambda < \infty. \tag{1.5.20}$$

Necessity. We calculate the following value

$$\begin{aligned}
\left\langle S e^{-i\lambda x}, e^{-i\lambda x} \right\rangle &= \int_0^\omega \frac{d}{dx} \int_0^\omega e^{-i\lambda t} s(x-t) \, dt \, e^{i\lambda x} \, dx \\
&= \int_0^\omega e^{-i\lambda t} s(\omega - t) \, dt \, e^{i\lambda \omega} \\
&\quad - \int_0^\omega e^{-i\lambda t} s(-t) \, dt - i\lambda \int_0^\omega \int_0^\omega e^{i\lambda(x-t)} s(x-t) \, dt \, dx.
\end{aligned}$$

Substituting in the last integral $u = x - t$ and changing the order of integration we obtain

$$\left\langle S e^{-i\lambda x}, e^{-i\lambda x} \right\rangle = \int_{-\omega}^\omega e^{i\lambda u} s(u) \, \operatorname{sgn} u \, du - i\lambda \int_{-\omega}^\omega e^{i\lambda u} s(u) \left(\omega - |u| \right) du. \tag{1.5.21}$$

Since

$$\left| \left\langle S e^{-i\lambda x}, e^{-i\lambda x} \right\rangle \right| \leq \|S\| \omega, \tag{1.5.22}$$

it follows from (1.5.19), (1.5.21) that

$$\left|\lambda\widetilde{S}_1(\lambda)\right| \le \|S\| + \frac{1}{\omega} \int\limits_{-\omega}^{\omega} |s(u)|\, du.$$

This proves the theorem. □

Corollary 1.5.3. *If $s(x)$ is a function of a bounded variation on the segment $[-\omega, \omega]$, then the corresponding operator S of the form* (1.0.1) *is bounded in $L^2(0,\omega)$.*

Indeed, it is known from the theory of trigonometric series that the function $\lambda\widetilde{S}(\lambda)$ is bounded on the axis $-\infty < \lambda < \infty$, if $s(x)$ has a bounded variation.

Corollary 1.5.4. *Let $s(x) \in L^2(-\omega, \omega)$, $xs'(x) \in L(-\omega, \omega)$. Then the requirement that the function $\lambda\widetilde{S}(\lambda)$ is uniformly bounded on the axis $-\infty < \lambda < \infty$ is necessary and sufficient for the boundedness of the operator S in the space $L^2(0,\omega)$.*

Proof. Sufficiency follows from Theorem 1.5.2.

To prove *necessity* we consider the integral

$$i\lambda \int\limits_{-\omega}^{\omega} e^{i\lambda u} s(u)|u|\, du = e^{i\lambda u} s(u)|u| \Big|_{-\omega}^{\omega} - \int\limits_{-\omega}^{\omega} e^{i\lambda u}\left(s'(u)|u| + s(u)\,\mathrm{sgn}\,u\right) du.$$

Now (1.5.21) can be written in the form

$$\left\langle Se^{-i\lambda x}, e^{-i\lambda x} \right\rangle = e^{i\lambda u} s(u)|u| \Big|_{-\omega}^{\omega} - i\lambda\omega\widetilde{S}(\lambda) - \int\limits_{-\omega}^{\omega} e^{i\lambda u} s'(u)|u|\, du.$$

Hence, by (1.5.22) we have

$$\left|\lambda\widetilde{S}(\lambda)\right| \le |s(\omega)| + |s(-\omega)| + \|S\| + \frac{1}{\omega}\int\limits_{-\omega}^{\omega} |s'(u)|\,|u|\, du. \tag{1.5.23}$$

This proves the corollary. □

Example 1.5.5. We consider the operator

$$Sf = \frac{\mathrm{d}}{\mathrm{d}x} \int\limits_{0}^{x} f(t)\ln|x - t|\, dt. \tag{1.5.24}$$

In this case

$$s(x) = \ln x, \quad x > 0; \qquad s(x) = 0, \quad x < 0.$$

The kernel $s(x)$ satisfies the conditions of Corollary 1.5.4. Integrating by parts we have

$$i\lambda \int_0^\omega e^{i\lambda t} \ln t \, dt = \left(e^{i\lambda\omega} - 1\right) \ln\omega - \int_0^\omega \left(e^{i\lambda t} - \mathbb{1}\right) \frac{dt}{t}$$

$$= \left(e^{i\lambda\omega} - 1\right) \ln\omega - \int_0^{\lambda\omega} \left(e^{ix} - \mathbb{1}\right) \frac{dx}{x}, \quad \lambda > 0. \qquad (1.5.25)$$

Since

$$\lim_{\lambda\to\infty} \int_0^{\lambda\omega} \frac{\mathbb{1} - \cos x}{x} \, dx = \infty,$$

from (1.5.25) it follows that $\lambda\widetilde{S}(\lambda)$ on the axis $-\infty < \lambda < \infty$ is unbounded. It means that according to Corollary 1.5.4, the operator S defined by (1.5.24) is unbounded in the space $L^2(0,\omega)$.

Chapter 2

Equations of the First Kind with a Difference Kernel

In this chapter we consider again the equation

$$Sf = \frac{\mathrm{d}}{\mathrm{d}x} \int_0^\omega f(t)s(x-t)\,\mathrm{d}t = \varphi(x). \qquad (2.0.1)$$

The operator S is now not assumed to be invertible. An important particular case of (2.0.1) is the equation of the first kind

$$Sf = \int_0^\omega f(t)k(x-t)\,\mathrm{d}t = \varphi(x), \quad k(x) \in L(-\omega,\omega). \qquad (2.0.2)$$

Equations of the form (2.0.2) play an essential role in a number of applied problems [101]. There are, however, few rigorous results in the theory of such equations. The regularization method is not adaptable to a calculation of the difference structure of the kernel of (2.0.1) and (2.0.2). In this chapter we develop a method that takes into account the specific character of the kernels of (2.0.1) and (2.0.2). As in Chapter 1, a solution of (2.0.1) can be expressed in terms of the particular solutions

$$SN_1 = M, \quad SN_2 = 1. \qquad (2.0.3)$$

The general results are illustrated and made more precise in the classical examples when the kernel $k(x)$ in (2.0.2) is one of the following functions:

$$\ln\frac{A}{2|x|}, \quad |x|^{-h} \ (-1 < h < 1), \quad \mathrm{e}^{-v|x|} \ (v > 0). \qquad (2.0.4)$$

Operators with kernels (2.0.4), or similar ones, have been the object of study of a number of authors (Carleman [12], Kreĭn [38], Gakhov [17], Zadeh and Ragazzini [113]).

2.1 Equations of the first kind with a special right-hand side

1. Let S be a bounded operator of the form

$$Sf = \frac{\mathrm{d}}{\mathrm{d}x} \int\limits_0^\omega f(t)s(x-t)\,\mathrm{d}t, \quad s(x) \in L^p(-\omega,\omega), \tag{2.1.1}$$

in the space $L^p(0,\omega)$ $(1 \le p \le 2)$, where $1/p + 1/q = 1$. The main equalities from Chapter 1 for S and S^* acting in $L^2(0,\omega)$ are also valid for S and S^* acting in $L^p(0,\omega)$. We note that the operators

$$Af = \mathrm{i}\int\limits_0^x f(t)\,\mathrm{d}t, \quad A^*f = -\mathrm{i}\int\limits_x^\omega f(t)\,\mathrm{d}t \tag{2.1.2}$$

are considered in $L^p(0,\omega)$ $(1 \le p \le 2)$ and, therefore, when $1 \le p < 2$ they are not adjoint to each other.

Proposition 2.1.1. *Let S be a bounded operator of the form* (2.1.1) *acting in* $L^p(0,\omega)$. *Then the operator identity* (1.1.10), *where M and N are given in* (1.1.11), *is valid. Formulas* (1.2.2) *and* (1.2.3) *hold for $\mathcal{L}_m$ given by* (1.2.1). *The operator S is bounded in all the spaces $L^r(0,\omega)$ $(p \le r \le q)$ and the adjoint operator S^* acting in the spaces $L^{\tilde r}(0,\omega)$ $(\tilde r = r/(r-1))$ is given by the equality*

$$S^* = USU, \quad Uf = \overline{f(\omega - x)}. \tag{2.1.3}$$

The statements of the proposition above are proved similar to the corresponding statements in Chapter 1 and we prove here only the following lemma.

Lemma 2.1.2. *If $f \in L^p(0,\omega)$ and $Sf \in L^q(0,\omega)$, then*

$$\langle Sf, Uf \rangle = \langle f, S^*Uf \rangle. \tag{2.1.4}$$

Proof. Since $Sf \in L^q(0,\omega)$, it follows from (2.1.3) that $S^*Uf \in L^q(0,\omega)$. Hence, both sides of (2.1.4) are well defined. Now (2.1.4) is obtained directly from the following property of an involution:

$$\langle f_1, f_2 \rangle = \langle Uf_2, Uf_1 \rangle, \quad f_1 \in L^q(0,\omega), \quad f_2 \in L^p(0,\omega). \tag{2.1.5}$$

$\square$

2. Theorem 1.3.1 from Chapter 1 can be modified for the case of the equations of the first kind as follows.

Theorem 2.1.3. *Suppose that an operator S of the form (2.1.1) is bounded in $L^p(0,\omega)$ $(1 \le p \le 2)$ and that there are functions $N_1(x)$ and $N_2(x)$ in $L^p(0,\omega)$ satisfying (2.0.3). Then*

$$SB_\gamma(x,\lambda) = e^{ix\lambda}, \quad \lambda \ne -\frac{1}{i\gamma}, \tag{2.1.6}$$

where

$$\gamma = \langle SN_1, UN_2 \rangle - \langle N_1, S^*UN_2 \rangle, \quad B_\gamma(x,\lambda) = \frac{1}{i\lambda\gamma + 1}B(x,\lambda), \tag{2.1.7}$$

and $B(x,\lambda)$ belongs to $L^p(0,\omega)$ and is defined by (1.3.4)–(1.3.6).

Proof. According to Proposition 2.1.1, the identity (1.1.10) and the recurrence formula (1.2.3) remain valid for S acting in $L^p(0,\omega)$. Since $N(x) \in L^q(0,\omega)$ and (1.2.3) holds, it follows that for a certain C we have

$$\|\mathcal{L}_{m+1}\|_p \le Cm\|\mathcal{L}_m\|_p \le C^{m+1}m! \tag{2.1.8}$$

Here $\|f\|_p$ is the norm in the space $L^p(0,\omega)$. By virtue of (2.1.8) the series

$$B_\gamma(x,\lambda) = \sum_{m=0}^{\infty} \frac{(i\lambda)^m}{M!}\mathcal{L}_{m+1}$$

converges for $|\lambda| < C^{-1}$. Consequently,

$$SB_\gamma(x,\lambda) = e^{i\lambda x}, \quad |\lambda| < C^{-1}. \tag{2.1.9}$$

Using (1.2.2) we deduce that

$$B_\gamma(x,\lambda) = u_\gamma(x,\lambda) - i\lambda \int_x^\omega B_\gamma(t,\lambda)\,dt, \tag{2.1.10}$$

where

$$u_\gamma(x,\lambda) = a_\gamma(\lambda)N_1(x) + b_\gamma(\lambda)N_2(x), \tag{2.1.11}$$

$$a_\gamma(\lambda) = i\lambda \int_0^\omega B_\gamma(t,\lambda)\,dt, \quad b_\gamma(\lambda) = 1 + i\lambda \int_0^\omega B_\gamma(t,\lambda)N(t)\,dt. \tag{2.1.12}$$

Solving the integral equation (2.1.10) we derive

$$B_\gamma(x,\lambda) = u_\gamma(x,\lambda) - i\lambda \int_x^\omega e^{i(x-t)\lambda}u_\gamma(t,\lambda)\,dt. \tag{2.1.13}$$

Next we write the functions $a_\gamma(\lambda)$ and $a(\lambda)$ in the following form:

$$a_\gamma(\lambda) = i\lambda \langle B_\gamma(x,\lambda), S^* U N_2 \rangle, \quad a(\lambda) = i\lambda \langle S B_\gamma(x,\lambda), U N_2 \rangle. \tag{2.1.14}$$

Then from (2.1.4), (2.1.11), and (2.1.13), we obtain

$$a(\lambda) = a_\gamma(\lambda) \left(1 + i\lambda\gamma \right), \tag{2.1.15}$$

where γ is defined by the first formula (2.1.7).

Writing $b_\gamma(\lambda)$ and $b(\lambda)$ in the form

$$b_\gamma(\lambda) = 1 + i\lambda \Big\langle B_\gamma, S^* U (\mathbb{1} - N_1) \Big\rangle, \quad b(\lambda) = 1 + i\lambda \Big\langle S B_\gamma, U (\mathbb{1} - N_1) \Big\rangle,$$

we derive a relation analogous to (2.1.15):

$$b(\lambda) = b_\gamma(\lambda) \left(1 + i\lambda\gamma \right). \tag{2.1.16}$$

From (2.1.15) and (2.1.16) it follows that

$$u_\gamma(x,\lambda) = \frac{u(x,\lambda)}{1 + i\lambda\gamma}, \quad B_\gamma(x,\lambda) = \frac{B(x,\lambda)}{1 + i\lambda\gamma}, \quad |\lambda| < C^{-1},$$

that is, the theorem is true for $|\lambda| < C^{-1}$. Since $B(x,\lambda)$ and $e^{i\lambda x}$ are analytic in λ, the theorem follows. $\square$

We introduce the function

$$\rho(\lambda,\mu) = \int_0^\omega B_\gamma(x,\lambda) e^{i\mu x} \, dx. \tag{2.1.17}$$

A formula analogous to (1.3.12) can be deduced from Theorem 2.1.3:

$$\rho(\lambda,\mu) = -\frac{i e^{i\omega\mu}}{i\lambda\gamma + 1} \frac{a(\lambda)b(-\mu) - b(\lambda)a(-\mu)}{\lambda + \mu}. \tag{2.1.18}$$

2.2   Solutions of equations of the first kind

Let S be an operator with a difference kernel that is bounded in $L^p(0,\omega)$ and suppose that there are functions $N_1(x)$ and $N_2(x)$ satisfying (2.0.3). A solution of (2.0.1) for the particular right-hand side $\varphi(x) = e^{i\lambda x}$ was given in Theorem 2.1.3. Using this result we construct a solution of (2.0.1) for the class $\varphi(x)$ from $W_p^{(2)}$.

1. We introduce the function

$$r(x,t) = N_2(\omega - t)N_1(x) - N_1(x - t)N_2(x). \tag{2.2.1}$$

From

$$r(x,t) = -r(\omega - t, \omega - x) \tag{2.2.2}$$

there follows the relation

$$\int\limits_x^\omega \int\limits_0^{\omega-x} f(x - t + s)r(t, s)\, ds\, dt = 0, \tag{2.2.3}$$

where $f(x)$ is an arbitrary function in $L^q(-\omega, \omega)$.

We denote by $W_p^{(l)}$ the set of functions $\varphi(x)$ such that $\varphi^{(l)}(x) \in L^p(0, \omega)$. We define an operator T on $W_p^{(2)}$ by

$$T\varphi = \int\limits_0^\omega \varphi'(t)r(x,t)\, dt + \varphi(\omega)N_2(x) - \int\limits_x^\omega \varphi'(x - t + \omega)N_2(t)\, dt$$

$$- \int\limits_x^\omega \int\limits_{\omega-x}^\omega \varphi''(x - t + s)r(t, s)\, ds\, dt. \tag{2.2.4}$$

It is easy to see that T maps $W_p^{(2)}$ into $L^p(0, \omega)$. Taking into account (2.2.3), from (1.3.4)–(1.3.6) we obtain

$$B(x, \lambda) = Te^{i\lambda x}. \tag{2.2.5}$$

If $\gamma = 0$ then (2.1.6) and (2.2.5) imply that

$$STe^{i\lambda x} = e^{i x \lambda}, \quad TSB(x, \lambda) = B(x, \lambda). \tag{2.2.6}$$

Using the first of the relations (2.2.6) we prove the following theorem.

Theorem 2.2.1. *Suppose that the conditions of Theorem 2.1.3 are fulfilled and that* $\gamma = 0$. *Then the operator T defined by (2.2.4) is a right inverse of S, that is,*

$$ST\varphi = \varphi, \quad \varphi \in W_p^{(2)}. \tag{2.2.7}$$

Thus, for $\gamma = 0$ *and* $\varphi \in W_p^{(2)}$ *the function* $f(x) = T\varphi$ *is a solution of (2.0.1).*

2. We introduce a simpler formula for T. To do this we need the identity

$$\int\limits_x^\omega \int\limits_{t-x}^\omega f(x - t + s)r(t, s)\, ds\, dt = \int\limits_x^\omega \int\limits_{\omega-x}^\omega f(x - t + s)r(t, s)\, ds\, dt, \tag{2.2.8}$$

which follows from (2.2.3) if we put $f(u) = 0$ for $-\omega < u < 0$. Further, in view of (2.2.2) we get

$$\int_x^\omega r(t, t - x)\, dt = 0. \qquad (2.2.9)$$

Taking (2.2.8) and (2.2.9) into account we rewrite (2.2.4) in the following form:

$$T\varphi = -\frac{d}{dx}\int_x^\omega\int_{t-x}^\omega \varphi'(x - t + s)r(t, s)\, ds\, dt - \frac{d}{dx}\int_x^\omega \varphi(x - t + \omega)N_2(t)\, dt. \quad (2.2.10)$$

Substituting the variables $s = u + t - x$ and $t = \dfrac{v + x - u}{2}$ we obtain the form

$$T\varphi = -\frac{1}{2}\frac{d}{dx}\int_0^\omega\left(\int_{x+u}^{2\omega-|x-u|} r\left(\frac{v + x - u}{2}, \frac{v + u - x}{2}\right) dv\right)\varphi'(u)\, du$$

$$-\frac{d}{dx}\int_x^\omega N_2(t)\varphi(x - t + \omega)\, dt. \qquad (2.2.11)$$

As in Chapter 1 we put

$$Q(x, t) = N_2(\omega - t)N_1(x) + \left(1 - N_1(\omega - t)\right)N_2(x),$$

$$\Phi(x, t) = \frac{1}{2}\int_{x+t}^{2\omega-|x-t|} Q\left(\frac{s + x - t}{2}, \frac{s - x + t}{2}\right) ds. \qquad (2.2.12)$$

Then the equality

$$Q(x, t) = r(x, t) + N_2(x) \qquad (2.2.13)$$

holds. A direct calculation shows that

$$\frac{1}{2}\frac{d}{dx}\int_0^\omega\left(\int_{x+u}^{2\omega-|x-u|} N_2\left(\frac{v + u - x}{2}\right) dv\right)\varphi'(u)\, du - \frac{d}{dx}\int_x^\omega N_2(t)\varphi(x - t + \omega)\, dt$$

$$= \varphi(0)N_2(x). \qquad (2.2.14)$$

From (2.2.11), (2.2.13)–(2.2.14) we deduce the final formula for T:

$$T\varphi = -\frac{d}{dx}\int_0^\omega \Phi(x, t)\varphi'(t)\, dt + \varphi(0)N_2(x). \qquad (2.2.15)$$

3. Now, let us consider the case where

$$\gamma = \langle SN_1, UN_2 \rangle - \langle N_1, S^*UN_2 \rangle \neq 0. \tag{2.2.16}$$

We put $\lambda_0 = -1/i\gamma$. Comparing the residues of both sides of (2.1.6) at λ_0, we see that

$$SB(x, \lambda_0) = 0. \tag{2.2.17}$$

We rewrite (2.1.6) as follows:

$$S\Big(B(x, \lambda) - B(x, \lambda_0)\Big)\Big/\Big(i\lambda\gamma + 1\Big) = e^{i\lambda x}. \tag{2.2.18}$$

We introduce the operators

$$C_\gamma\varphi = \frac{1}{\gamma}\int_0^x \varphi(t)e^{i\lambda_0(x-t)}\,\mathrm{d}t, \quad T_\gamma = TC_\gamma. \tag{2.2.19}$$

It is easy to see that

$$C_\gamma e^{i\lambda x} = \frac{e^{i\lambda x} - e^{i\lambda_0 x}}{i\lambda\gamma + 1}. \tag{2.2.20}$$

By (2.2.5), (2.2.19), and (2.2.20),

$$\frac{B(x, \lambda) - B(x, \lambda_0)}{i\lambda\gamma + 1} = T_\gamma e^{i\lambda x}. \tag{2.2.21}$$

From (2.2.15) and (2.2.19) we see that T maps $W_p^{(1)}$ into $L^p(0, \omega)$ and can be written in the form

$$T_\gamma\varphi = -\frac{1}{\gamma}\frac{\mathrm{d}}{\mathrm{d}x}\int_0^\omega \left(\varphi(t) + i\lambda_0\int_0^t \varphi(u)e^{i\lambda_0(t-u)}\,\mathrm{d}u\right)\Phi(x, t)\,\mathrm{d}t. \tag{2.2.22}$$

The relations (2.2.18) and (2.2.21) show that $ST_\gamma e^{i\lambda x} = e^{i\lambda x}$. Thus, we have the following theorem.

Theorem 2.2.2. *Suppose that the conditions of Theorem 2.1.3 are fulfilled and that $\gamma \neq 0$. Then the operator T_γ defined by (2.2.22) is a right inverse of S, that is,*

$$ST_\gamma\varphi = \varphi, \quad \varphi(x) \in W_p^{(1)}. \tag{2.2.23}$$

4. The problem of constructing the operators T and T_γ, cf. (2.2.19), is simplified if

$$R = \int_0^\omega N_2(t)\,\mathrm{d}t \neq 0. \tag{2.2.24}$$

Then it follows from (1.2.2) for $m = 1$ that

$$N_1(x) = \frac{1}{R} \left(\int_x^\omega N_2(t)\,dt + \mathcal{L}_2(x) + \alpha N_2(x) \right), \qquad \alpha = - \int_0^\omega N_2(t) N(t)\,dt.$$

(2.2.25)

When we now put

$$Q_1(x,t) = \frac{1}{R} \Big(N_2(\omega - t)\mathcal{L}_2(x) - \mathcal{L}_2(\omega - t) N_2(x) \Big),$$

(2.2.26)

$$Q_2(x,t) = \frac{1}{R} \left(N_2(\omega - t) \int_x^\omega N_2(s)\,ds + \int_0^{\omega - t} N_2(s)\,ds\, N_2(x) \right),$$

(2.2.27)

we deduce from (2.2.12) and (2.2.25)–(2.2.27) that

$$Q(x,t) = Q_1(x,t) + Q_2(x,t).$$

(2.2.28)

The corresponding function $\Phi(x,t)$ has the form

$$\Phi(x,t) = \Phi_1(x,t) + \Phi_2(x,t),$$

(2.2.29)

where

$$\Phi_k(x,t) = \frac{1}{2} \int_{x+t}^{2\omega - |x-t|} Q_k \left(\frac{s + x - t}{2}, \frac{s - x + t}{2} \right) ds \quad (k = 1, 2).$$

(2.2.30)

Using (2.2.15), (2.2.29), and (2.2.30) we represent T as follows:

$$T\varphi = -\frac{d}{dx} \int_0^\omega \varphi'(t)\Phi_1(x,t)\,dt + \frac{1}{R} \int_0^\omega \varphi(t) N_2(\omega - t)\,dt\, N_2(x).$$

(2.2.31)

Thus, under the condition (2.2.24), it is enough to know the two functions $\mathcal{L}_1(x)$ and $\mathcal{L}_2(x)$ in order to construct the operator T.

It follows from (2.2.24) and (2.2.25) that formula (2.1.7) for γ can be written in the form

$$\gamma = \frac{1}{R} \int_0^\omega \Big(t\mathcal{L}_1(\omega - t) - \mathcal{L}_2(t) \Big)\,dt.$$

(2.2.32)

5. We note that (2.2.26)–(2.2.31) also hold when S has a bounded inverse T in $L^2(0, \omega)$. In this case T admits the representation

$$T\varphi = \frac{d}{dx} \int_0^\omega \left(\frac{\partial}{\partial t} \Phi_1(x,t) \right) \varphi(t)\,dt + \frac{1}{R} \int_0^\omega N_2(\omega - t)\varphi(t)\,dt\, N_2(x).$$

(2.2.33)

6. Under certain conditions, Theorems 2.2.1 and 2.2.2 provide a procedure to construct solutions of (2.0.1). The description of the whole set of solutions and also the question of uniqueness of the solution leads, as usual, to the study of the equation

$$Sf = 0. \qquad (2.2.34)$$

Here H_S stands for the set of solutions of (2.2.34) belonging to $L^p(0,\omega)$. If $\dim H_S > 1$, then the following theorem is useful.

Theorem 2.2.3. *Let S be an operator with a difference kernel that is bounded in $L^p(0,\omega)$ ($1 \leq p \leq 2$) and $1 < \dim H_S \leq n < \infty$. Then H_S has a basis f_k ($0 \leq k \leq n-1$) such that*

$$f_{k+1} = A^* f_k, \quad 0 \leq k \leq n-2, \qquad (2.2.35)$$

where A and A^* are defined by (2.1.2).

Proof. Since the operator A^* has no finite-dimensional invariant subspaces, the inequality $A^* H_S \neq H_S$ is valid. According to (1.1.10), the operator S maps $A^* H_S$ into the subspace spanned on $M(x)$ and $\mathbb{1}$. This subspace cannot be two-dimensional, because in that case there would exist absolutely continuous functions N_1 and N_2, and by virtue of Theorem 1.4.3, S would be invertible. Thus, $\dim SA^* H_S = 1$. Hence, the subspace $A^* H_S$ has a common part H_S^1 of dimension $n-1$ with H_S. Similarly we derive that

$$A^* H_S \neq H_S^1, \quad \dim SA^* H_S^1 \leq 1. \qquad (2.2.36)$$

Putting $H_S^2 = A^* H_S^1 \cap H_S$, we deduce from (2.2.36) that

$$\dim H_S^2 \geq n-2. \qquad (2.2.37)$$

It is easy to see that $H_S^2 \subset H_S^1$, and so we have the equality in (2.2.37). Repeating this process we obtain subspaces $H_S^k = A^* H_S^{k-1} \cap H_S$ ($2 \leq k \leq n-1$), and

$$H_S \supset H_S^1 \supset \cdots \supset H_S^{n-1}, \quad \dim H_S^k = n-k. \qquad (2.2.38)$$

Thus, there exists a function $f_0 \in H_S$ such that

$$f_k = A^{*k} f_0 \in H_S^k \;\; (1 \leq k \leq n-1), \quad \|f_0\|_p \neq 0. \qquad (2.2.39)$$

The system of functions $f_1, f_2, \ldots, f_{n-1}$ is linearly independent, because from

$$\left(\alpha_0 + \alpha_1 A^* + \cdots + \alpha_{n-1} A^{*n-1}\right) f_0 = 0, \qquad (2.2.40)$$

it follows that $\alpha_0 = \alpha_1 = \cdots = \alpha_{n-1} = 0$. Thus, the functions f_k ($0 \leq k \leq n-1$) form a basis of H_S and satisfy (2.2.35). This proves the theorem. $\qquad \square$

A similar fact was proved by M.G. Kreĭn (verbal communication) for operators of the form

$$Sf = f(x) + \int_0^\omega f(t) k(\omega - t)\, dt, \quad k(x) \in L(-\omega, \omega). \qquad (2.2.41)$$

7. Under the assumption that $N_1(x)$ and $N_2(x)$ exist, the following theorem is true.

Theorem 2.2.4. *Suppose that an operator S of the form* (2.1.1) *is bounded in $L^p(0, \omega)$ and that there are functions $N_1(x)$ and $N_2(x)$ in $L^p(0, \omega)$ satisfying* (2.0.3).

   I. *If $f \in H_S$ and $\|f\|_p \neq 0$, then $f \notin L^q(0, \omega)$.*
  II. *If $p = 2$, then $\dim H_S = 0$.*
 III. *The inequality $\dim H_S \leq 1$ is valid.*
  IV. *If $\gamma \neq 1$ then $\dim H_S = 1$.*

Proof. Assume that $f \in L^q(0, \omega)$. By Theorem 2.1.3,

$$\left\langle e^{ix\lambda}, Uf \right\rangle = \left\langle SB_\gamma(x, \lambda), Uf \right\rangle. \tag{2.2.42}$$

Using (2.1.3) we deduce from (2.2.42) that

$$\left\langle e^{ix\lambda}, Uf \right\rangle = 0,$$

that is, $f(x) = 0$, which contradicts the condition $\|f\|_p \neq 0$. This proves the first assertion. Assertion II is immediate from the first assertion. It also follows from the first assertion that $A^* H_S$ and H_S do not have common non-trivial elements. On the other hand, similar to the proof of Theorem 2.2.3, we deduce that

$$\dim SA^* H_S \leq 1.$$

Hence, we have $\dim A^* H_S \leq 1$, from which we derive III.

Now we consider the case, where $\gamma \neq 0$. We prove that

$$\|B(x, \lambda_0)\|_p \neq 0. \tag{2.2.43}$$

Let us assume that

$$\|B(x, \lambda_0)\|_p = 0. \tag{2.2.44}$$

From (1.3.4), (1.3.5) and (2.2.44) it follows that

$$a(\lambda_0)N_1(x) + b(\lambda_0)N_2(x) = 0. \tag{2.2.45}$$

From (1.3.6) and (2.2.45) we deduce, that

$$a(\lambda_0) = b(\lambda_0) = 0. \tag{2.2.46}$$

Putting

$$u_0(x) = a'(\lambda_0)N_1(x) + b'(\lambda_0)N_2(x), \tag{2.2.47}$$

$$B_0(x) = u_0(x) - i\lambda_0 \int_x^\omega e^{i\lambda_0(x-t)} u_0(t)\, dt, \tag{2.2.48}$$

we deduce from (2.1.6) that

$$\frac{1}{i\gamma} SB_0(x) = e^{i\lambda_0 x}.$$ (2.2.49)

Using the notation (2.1.17) and formula (2.2.49), we have

$$\frac{1}{i\gamma}\Big\langle SB_0(x), UB_\gamma(x, \lambda)\Big\rangle = e^{i\omega\lambda}\rho(\lambda, -\lambda_0).$$ (2.2.50)

Now, in view of (2.2.46), (2.2.50), and (2.1.18), we obtain

$$\Big\langle SB_0(x), UB_\gamma(x, \lambda)\Big\rangle = 0.$$

Hence, taking into account (2.1.4), we see that

$$\Big(a'(\lambda_0)b(\lambda) - b'(\lambda_0)a(\lambda)\Big)\gamma + \Big\langle B_0(x), S^*UB_\gamma(x, \lambda)\Big\rangle = 0.$$ (2.2.51)

Formulas (2.1.3) and (2.1.6) yield

$$\Big\langle B_0(x), S^*UB_\gamma(x, \lambda)\Big\rangle = \int_0^\omega B_0(x)e^{i(\omega-x)\lambda}\, dx.$$ (2.2.52)

The formulas (2.2.48) and (2.2.52) lead to

$$\Big\langle B_0(x), S^*UB_\gamma(x, \lambda)\Big\rangle = \frac{a'(\lambda_0)\Big(b(\lambda) + e^{i\lambda_0\omega} - e^{i\lambda\omega}\Big) - b'(\lambda_0)a(\lambda)}{i(\lambda_0 - \lambda)}.$$ (2.2.53)

We substitute the right-hand side of (2.2.53) in (2.2.51):

$$\Big(a'(\lambda_0)b(\lambda) - b'(\lambda_0)a(\lambda)\Big)\left(\gamma + \frac{1}{i(\lambda_0 - \lambda)}\right) + a'(\lambda_0)\frac{e^{i\lambda_0\omega} - e^{i\lambda\omega}}{i(\lambda_0 - \lambda)} = 0.$$ (2.2.54)

Taking (2.2.46) into account we pass in (2.2.54) to the limit as $\lambda \to \lambda_0$:

$$a'(\lambda_0) = 0.$$ (2.2.55)

It follows from (2.2.54) and (2.2.55) that

$$b'(\lambda_0)a(\lambda)\left(\gamma\frac{1}{i(\lambda_0 - \lambda)}\right) = 0.$$

Hence,

$$b'(\lambda_0) = 0.$$ (2.2.56)

The equations (2.2.55) and (2.2.56) indicate that $\|u_0(x)\|_p = 0$. Then, by (2.2.47), $\|B_0(x)\|_p = 0$, which contradicts (2.2.49). Hence, (2.2.44) does not hold and $\|B(x, \lambda_0)\|_p \neq 0$. Now IV follows from III and (2.2.17). This proves the theorem. □

8. The following result is useful for the study of various concrete examples.

Theorem 2.2.5. *Suppose that an operator S is defined by (2.0.2) and that there are functions $N_1(x)$ and $N_2(x)$ in $L^p(0, \omega)$ $(1 \le p \le 2)$ satisfying (2.0.3). If the functions*

$$g_m(x) = \int_0^\omega |N_m(t)| \cdot |k(x - t)| \, dt \quad (m = 1, 2)$$

belong to $L^q(0, \omega)$, then $\gamma = 0$ and $\dim H_S = 0$.

Proof. The operator S^* has the form

$$S^* f = \int_0^\omega f(t) \overline{k(t - x)} \, dt.$$

Then

$$\left\langle N_1, S^* U N_2 \right\rangle = \int_0^\omega N_1(x) \int_0^\omega N_2(\omega - t) k(t - x) \, dt \, dx. \qquad (2.2.57)$$

Under the conditions of the theorem, the integral on the right-hand side of (2.2.57) converges absolutely. Hence, by Fubini's theorem we can change the order of integration:

$$\left\langle N_1, S^* U N_2 \right\rangle = \int_0^\omega \int_0^\omega N_1(x) k(t - x) \, dx \, N_2(\omega - t) \, dt = \left\langle S N_1, U N_2 \right\rangle. \qquad (2.2.58)$$

By (2.1.7) and (2.2.58),

$$\gamma = 0. \qquad (2.2.59)$$

Similarly,

$$\left\langle S N_m, U f \right\rangle = \left\langle N_m, S^* U f \right\rangle \qquad (2.2.60)$$

if

$$f \in L^p(0, \omega), \quad S^* U f \in L^q(0, \omega).$$

From Theorem 2.1.3 we obtain

$$S B(x, \lambda) = e^{i\lambda x}. \qquad (2.2.61)$$

Using (2.2.59)–(2.2.61) we deduce that

$$\left\langle e^{i\lambda x}, U f \right\rangle = \left\langle S B(x, \lambda), U f \right\rangle = \left\langle B(x, \lambda), S^* U f \right\rangle. \qquad (2.2.62)$$

If $f \in H_S$ then $\left\langle e^{i\lambda x}, U f \right\rangle = 0$, by (2.1.3) and (2.2.62), that is, $\|f\|_p = 0$. This proves the theorem. $\qquad \square$

2.3 Generalized solutions

In many cases, equation (2.0.2) has only generalized functions as solutions. This situation is typical for equations connected with optimal problems of automatic control [101]. Under certain assumptions the results of Sections 2.1, 2.2 can be extended to this case.

1. We denote by $\mathfrak{D}$ the set of generalized functions of the form

$$f(x) = \alpha\delta(x) + \beta\delta(\omega - x) + f_1(x), \tag{2.3.1}$$

where $f_1(x) \in L(0, \omega)$ and $\delta(x)$ is the delta function.

We say that $g(x)$ belongs to the basic space K, if $g(x)$ is bounded on $[0, \omega]$ and is continuous at 0 and ω.

As usual (see [18, Chap. 1]) a generalized function f is a linear functional on K:

$$\langle g, f \rangle = \overline{\langle f, g \rangle} = \int_0^\omega g(x)\overline{f(x)}\,\mathrm{d}x. \tag{2.3.2}$$

By definition,

$$\langle g, f \rangle = \overline{\alpha}g(0) + \overline{\beta}g(\omega) + \int_0^\omega g(x)\overline{f}_1(x)\,\mathrm{d}x. \tag{2.3.3}$$

We introduce the operator

$$Sf = \int_0^\omega f(t)k(x - t)\,\mathrm{d}t, \tag{2.3.4}$$

where $k(x)$ is a continuous function on $[-\omega, \omega]$. The operator S maps functions from $\mathfrak{D}$ into continuous functions on $[0, \omega]$. Operators A and A^* on $\mathfrak{D}$ are defined by

$$Af = \mathrm{i}\int_0^x f(t)\,\mathrm{d}t = \mathrm{i}\alpha + \mathrm{i}\int_0^x f_1(t)\,\mathrm{d}t, \tag{2.3.5}$$

$$A^*f = -\mathrm{i}\int_x^\omega f(t)\,\mathrm{d}t = -\mathrm{i}\beta - \mathrm{i}\int_x^\omega f_1(t)\,\mathrm{d}t. \tag{2.3.6}$$

Theorem 2.3.1. *For any function f in $\mathfrak{D}$*

$$(AS - SA^*)f = \mathrm{i}\int_0^\omega f(t)\Big(M(x) + N(t)\Big)\,\mathrm{d}t, \tag{2.3.7}$$

where

$$M(x) = \int_0^x k(t)\, dt, \quad N(x) = -\int_0^{-x} k(t)\, dt. \tag{2.3.8}$$

Proof. From (2.3.4)–(2.3.6) and (2.3.8) it follows that

$$\left(AS - SA^*\right)\delta(x) = i\int_0^x k(t)\, dt = i\int_0^\omega \left(M(x) + N(t)\right)\delta(t)\, dt, \tag{2.3.9}$$

$$\left(AS - SA^*\right)\delta(\omega - x) = i\int_0^x k(t - \omega)\, dt + i\int_0^\omega k(x - t)\, dt. \tag{2.3.10}$$

Comparing (2.3.8) and (2.3.10) we have

$$\left(AS - SA^*\right)\delta(\omega - x) = i\left(M(x) + N(\omega)\right) = i\int_0^\omega \left(M(x) + N(t)\right)\delta(\omega - t)\, dt. \tag{2.3.11}$$

In view of (2.3.9) and (2.3.11), the equality (2.3.7) holds for $f(x) = \delta(x)$ and $f(x) = \delta(\omega - x)$. The fact that (2.3.7) holds for $f(x) = f_1(x) \in L(0,\omega)$ is proved in a way which is similar to the proof of Theorem 1.1.3. Thus, (2.3.7) is valid for $f \in \mathfrak{D}$. This proves the theorem. □

Now, together with the operator S we consider the operator

$$S^* f = \int_0^\omega \overline{k(x - t)} f(t)\, dt \tag{2.3.12}$$

which maps the functions from $\mathfrak{D}$ into the functions which are continuous on the segment $[0,\omega]$. Let us prove that for any pair of the functions $f(x)$ and $g(x)$ from $\mathfrak{D}$ the following equality holds:

$$\langle Sf, g\rangle = \langle f, S^* g\rangle. \tag{2.3.13}$$

Indeed, we put

$$f(x) = \alpha\,\delta(x) + \beta\,\delta(\omega - x) + f_1(x), \quad g(x) = \gamma\,\delta(x) + \nu\,\delta(\omega - x) + g_1(x),$$

where $f_1(x), g_1(x) \in L(0,\omega)$. By (2.3.4) and (2.3.12) we obtain

$$Sf = \alpha k(x) + \beta k(x - \omega) + \int_0^\omega k(x - t) f_1(t)\, dt,$$

$$S^* g = \gamma \overline{k(-x)} + \nu \overline{k(\omega - x)} + \int_0^\omega \overline{k(t - x)} g_1(t)\, dt.$$

After some calculations we see that the left-hand side of (2.3.13) equals the right-hand side.

2. Theorem 2.3.1 and relations (2.3.13) allow us to modify the results from Section 2.1 for the present case.

Theorem 2.3.2. *Suppose that S has the form* (2.3.4) *and that there exist functions N_1 and N_2 satisfying* (2.0.3). *Then the function $B(x, \lambda)$ defined by* (1.3.4)–(1.3.6) *belongs to $\mathfrak{D}$ and*

$$SB(x, \lambda) = e^{i x \lambda}. \tag{2.3.14}$$

Proof. For a function f of the form (2.3.1) we introduce the norm

$$\|f\|_{\mathfrak{D}} = |\alpha| + |\beta| + \int_0^\omega |f_1(t)| \, dt.$$

Since $N(x)$ is bounded on $[0, \omega]$, it follows from (1.2.2) that for some c

$$\|\mathcal{L}_{m+1}\|_{\mathfrak{D}} \le cm\|\mathcal{L}_m\|_{\mathfrak{D}} \le c^{m+1} m!.$$

Hence, the series

$$B(x, \lambda) = \sum_{m=0}^{\infty} \frac{(i\lambda)^m}{m!} \mathcal{L}_{m+1}$$

converges for $|\lambda| < c^{-1}$. We also see that $B(x, \lambda) \in \mathfrak{D}$ and

$$SB(x, \lambda) = e^{i\lambda x}. \tag{2.3.15}$$

As in Theorem 1.3.1 we pass to the integral equation

$$B(x, \lambda) = u(x, \lambda) - i\lambda \int_x^\omega B(t, \lambda) \, dt, \tag{2.3.16}$$

where $u(x, \lambda)$ is defined by (1.3.5) and (1.3.6).

To solve (2.3.16) we use the rule for changing the order of integration:

$$\int_x^\omega \int_t^\omega f(v)g(v, t) \, dv \, dt = \int_x^\omega f(v) \int_x^v g(v, t) \, dt \, dv, \tag{2.3.17}$$

where $f(x) \in \mathfrak{D}$, and $g(x, t)$ is a continuous function of x and t ($0 \le x, t \le \omega$).

We easily check that (2.3.17) holds for $f(x) = \delta(x)$ and $f(x) = \delta(\omega - x)$. It is known that formula (2.3.17) holds also for $f(x) \in L(0, \omega)$. Taking into account (2.3.17) we solve (2.3.16) by the method of successive approximations. This completes the proof of our assertion. □

Now, we introduce the function

$$\rho(\lambda, \mu) = \int\limits_0^{\omega} B(x, \lambda) e^{i\mu x}\, dx. \qquad (2.3.18)$$

From Theorem 2.3.2 and (2.3.17) we derive that

$$\rho(\lambda, \mu) = -ie^{i\omega\mu} \frac{a(\lambda)b(-\mu) - b(\lambda)a(-\mu)}{\lambda + \mu}. \qquad (2.3.19)$$

3. The set of functions $\varphi(x)$, which have continuous second derivatives on $[0, \omega]$ is denoted by $C^{(2)}$. Like in Section 2.2, the operator T is introduced by formulas (2.2.1) and (2.2.4). This operator maps functions $\varphi(x)$ from $C^{(2)}$ into functions from $\mathfrak{D}$.

Theorem 2.3.3. *Suppose that the conditions of Theorem 2.3.2 are satisfied. If $\varphi \in C^{(2)}$, then the equation (of the first kind)*

$$Sf = \varphi \qquad (2.3.20)$$

has a unique solution $f(x) = T\varphi$ in $\mathfrak{D}$.

Proof. The relation (2.2.3) remains valid for $N_1, N_2 \in \mathfrak{D}$. Using Theorem 2.3.2 and formulas (2.2.1), (2.2.3), and (2.2.4) we deduce that $Te^{i\lambda x} = B(x, \lambda)$. Hence,

$$STe^{i\lambda x} = e^{i\lambda x}, \quad \text{that is,} \quad ST\varphi = \varphi, \quad \varphi \in C^{(2)}.$$

Thus, the generalized function $f(x) = T\varphi$ is a solution of (2.3.20). Suppose that this equation has more than one solution. Then there is a non-trivial solution of

$$Sf = 0. \qquad (2.3.21)$$

It follows from (2.3.12) and (2.3.21) that

$$S^* U f = 0, \quad \text{where} \quad Uf = \overline{f(\omega - x)}.$$

Using Theorem 2.3.2, we have

$$\langle e^{ix\lambda}, Uf \rangle = \langle SB(x, \lambda), Uf \rangle.$$

When we now take into account (2.3.10), we obtain

$$\langle e^{ix\lambda}, Uf \rangle = 0, \quad \text{that is,} \quad f = 0.$$

This proves the theorem. $\qquad \square$

4. We consider the equation

$$Sf = \int_0^\omega e^{-\nu|x-t|} f(t) \, dt = \varphi(x), \quad \nu > 0. \tag{2.3.22}$$

After direct substitution we see that

$$N_1(x) = -\frac{1}{\nu} \delta(x), \tag{2.3.23}$$

$$N_2(x) = \frac{1}{2} \left(\nu + \delta(x) + \delta(\omega - x) \right). \tag{2.3.24}$$

Thus, according to (2.2.1), (2.2.4), (2.3.23), (2.3.24), for $\varphi \in C^{(2)}$, the equation (2.3.22) has one and only one solution in $\mathfrak{D}$.

2.4 On the behavior of solutions

Let us consider a bounded operator

$$Sf = \frac{d}{dx} \int_0^\omega s(x-t) f(x) \, dt, \quad s(x) \in L^q(-\omega, \omega), \quad \frac{1}{p} + \frac{1}{q} = 1 \tag{2.4.1}$$

acting in the space $L^p(0, \omega)$ ($1 \le p \le 2$).

We study the equation

$$Sf = \varphi \tag{2.4.2}$$

and obtain conditions of the boundedness of $f(x)$ We also describe the behavior of $f(x)$ at the ends of the segment $[0, \omega]$. Such questions arise in a number of applied problems [1, 37, 101]. Separately we study the equations of the form (2.4.2), the solutions of which are in the class of generalized functions.

1. Let the conditions of Theorem 2.1.3 be fulfilled and the equation $Sf = 0$ has in $L^p(0, \omega)$ only one solution.

Then the equation (2.4.2) has one and only one solution

$$f(x) = T\varphi, \tag{2.4.3}$$

where the operator T is defined by (2.2.4).

We put

$$P_1\varphi = \int_0^\omega N_2(\omega - t)\varphi'(t) \, dt, \quad P_2(\varphi) = \varphi(\omega) - \int_0^\omega N_1(\omega - t)\varphi'(t) \, dt.$$

From (2.2.4) and (2.4.3) we have the assertions [60]:

Proposition 2.4.1. *The solution $f(x)$ of the equation (2.4.2) has the form*

$$f(x) = P_1(\varphi)N_1(x) + P_2(\varphi)N_2(x) + O(1), \quad 0 \leq x \leq w. \qquad (2.4.4)$$

Proposition 2.4.2. *If any non-trivial linear combination $N_1(x)$ and $N_2(x)$ is an unbounded function, then for the boundedness of the solution of (2.4.2) it is necessary and sufficient that the equations*

$$P_1(\varphi) = 0, \quad P_2(\varphi) = 0 \qquad (2.4.5)$$

be fulfilled.

Proposition 2.4.3. *If the relations (2.4.5) are fulfilled, then the solution of the equation (2.4.2) is a function continuous on the segment $[0, w]$ where*

$$f(0) = f(w) = 0. \qquad (2.4.6)$$

2. The length w of the segment $[0, w]$, on which the integral equation (2.4.2) is considered, is physically meaningful in a number of problems. For instance, in the contact theory of elasticity [1] the value w coincides with the length of the region of the contact. In the theory of optimum synthesis, w is the value of the apparatus memory [113]. Thus, the choice of w is determined by the corresponding physical requirements.

Problem 2.4.4. *Let the kernel $s(x)$ be defined on the segment $-\Omega \leq x \leq \Omega$ and let the set of the functions $\varphi_w(x)$ $(0 \leq x \leq w \leq \Omega)$ be given. Find such a value w that the solution of the equation*

$$Sf_w = \varphi_w(x), \quad 0 \leq x \leq w \qquad (2.4.7)$$

is continuous on the segment $[0, w]$, and

$$f_w(0) = f_w(w) = 0.$$

By Proposition 2.4.3 the solutions of Problem 2.4.4 coincide with the roots of the system of the transcendental equations

$$P_1(\varphi_w) = 0, \quad P_2(\varphi_w) = 0. \qquad (2.4.8)$$

Problem 2.4.4 is not always solvable. For example, for $\varphi_w(x) = 1$, the system (2.4.8) has no solution.

3. Let the invertible operator

$$Sf = f(x) + \int_0^w k(x - t)f(t)\, dt \qquad (2.4.9)$$

act in $L(0, \omega)$. Then the functions $N_1(x)$, $N_2(x)$ are absolutely continuous and (see Section 1.4)

$$N_1(\omega)N_2(0) - N_2(\omega)\Big(N_1(0) - 1\Big) = 1. \qquad (2.4.10)$$

From (2.4.10) it follows that $N_2(x)$ does not map into zero simultaneously at both ends of the segment $[0, \omega]$.

Using the absolute continuity of $N_1(x)$ and $N_2(x)$ we write the functionals $P_1(\varphi)$ and $P_2(\varphi)$ in the form

$$P_1(\varphi) = N_2(0)\varphi(\omega) - N_2(\omega)\varphi(0) + \int_0^\omega N_2'(t)\varphi(\omega - t)\, dt, \qquad (2.4.11)$$

$$P_2(\varphi) = \Big(1 - N_1(0)\Big)\varphi(\omega) + N_1(\omega)\varphi(0) - \int_0^\omega N_1'(t)\varphi(\omega - t)\, dt. \qquad (2.4.12)$$

In the case of the operator S in the form (2.4.9), it is possible to weaken the condition $\varphi'' \in L^p(0, \omega)$ replacing it by the demand of the continuity of $\varphi(x)$. In this case the following assertion holds (see (2.4.11), (2.4.12)).

Proposition 2.4.5. *Relation* (2.4.5) *is a necessary and sufficient condition under which* (2.4.6) *holds.*

4. Now we consider the equations

$$Sf = \int_0^\omega k(x - t)f(t)\, dt = \varphi(x), \quad \varphi(x) \in C^{(2)} \qquad (2.4.13)$$

where $k(x)$ is continuous on the segment $[-\omega, \omega]$. Let $N_1(x)$, $N_2(x) \in \mathfrak{D}$. Then by Theorem 2.3.3 the equation (2.4.13) has one and only one solution $f(x) = T\varphi$, where the operator T is defined by (2.2.4). Using (2.2.4) we obtain the following assertion.

Proposition 2.4.6. *Let* $N_1, N_2 \in \mathfrak{D}$ *and the equation* (2.4.5) *hold. Then the solution* $f(x)$ *of the equation* (2.4.13) *is a continuous function.*

5. In the theory of optimal synthesis the following problem is essential.

Problem 2.4.7. *Let the continuous kernel* $k(x)$ *be defined on the segment* $-\Omega \leq t \leq \Omega$ *and a set of the function* $\varphi_\omega(x)$ *be given. Find such a value* ω *that the solution of the equation*

$$Sf_\omega = \int_0^\omega k(x - t)f_\omega(t)\, dt = \varphi_\omega(x), \quad \varphi_\omega(x) \in C^{(2)}$$

is continuous on the segment $[0, \omega]$.

According to Proposition 2.4.6 the solutions of Problem 2.4.7 coincide with the roots of the system of equations (2.4.8). If $\varphi_\omega(x) = 1$, then Problem 2.4.7 has no solution.

2.5   On a class of integro-differential equations

1. Let the operator S act in $L(-a, a)$ and let it be defined by the equality

$$Sf = \mu f(x) + \int_{-a}^{a} k(x - t)f(t)\, dt \qquad (2.5.1)$$

and $k(x) \in L(-a, a)$. We introduce the operator

$$\mathcal{A}f = -\frac{d}{dx} S \frac{d}{dx} f, \qquad (2.5.2)$$

where $\mathcal{A}$ is acting in the domain $\mathfrak{D}_{\mathcal{A}}$ of the functions f satisfying the conditions

$$f''(x) \in L(-a, a), \quad f(-a) = f(a) = 0. \qquad (2.5.3)$$

A number of problems of the Lévy processes theory lead to the equations of the form (see [30, 33, 97])

$$\mathcal{A}f = \varphi, \quad \varphi \in L(-a, a). \qquad (2.5.4)$$

The integro-differential equations of this kind arise in some problems of diffraction theory [24]. This section is dedicated to the investigation of the equation (2.5.4), the solution of which, in fact, reduces to the inversion of the operator S with a difference kernel.

2. Let us show that the following equality (see [21, 100]) holds:

$$S\left(\frac{d}{dx} f\right) - \frac{d}{dx}(Sf) = k(x - a)f(a) - k(x + a)f(-a), \qquad (2.5.5)$$

where $f'(x) \in L(-a, a)$. Integrating by parts, we have

$$S\left(\frac{d}{dx} f\right) = \mu f'(x) + \int_{-a}^{a} k'(x - t)f(t)\, dt + k(x - a)f(a) - k(x - a)f(-a). \quad (2.5.6)$$

From (2.5.6) and the equality

$$\frac{d}{dx} Sf = \mu f'(x) + \int_{-a}^{a} k'(x - t)f(t)\, dt$$

it follows that (2.5.5) holds in the case of smooth $k(x)$.

We can prove that (2.5.5) is true in the general case approximating non-smooth kernels by smooth ones. If $f''(x) \in L(-a, a)$, then from (2.5.5) we have

$$S\frac{d^2 f}{dx^2} - \frac{d}{dx}S\frac{d}{dx}f = k(x-a)f'(a) - f'(-a)k(x+a). \tag{2.5.7}$$

From (2.5.7) we deduce that $\mathcal{A}f \in L(-a, a)$ when $f'' \in L(-a, a)$.

We say that the operator S of the form (2.5.1) is a regular one, if the following conditions are fulfilled:

1) The equation

$$Sf = 0 \tag{2.5.8}$$

has in $L(-a, a)$ only a trivial solution.

2) In $L(-a, a)$ there exist functions $\mathcal{L}_k(x)$ $(k = 1, 2)$ such that

$$S\mathcal{L}_k = x^{k-1}, \quad k = 1, 2 \tag{2.5.9}$$

and

$$R = \int\limits_{-a}^{a} \mathcal{L}_1(x)\, dx \neq 0. \tag{2.5.10}$$

We remark that when $\mu \neq 0$, the invertibility of the operator S follows from the condition of regularity. We introduce the operator

$$T\varphi = -\frac{d}{dx}\int\limits_{-a}^{a} \Phi(x, y)\varphi'(y)\, dy + \frac{1}{R}\int\limits_{-a}^{a} \mathcal{L}_1(-y)\varphi(y)\, dy \mathcal{L}_1(x), \tag{2.5.11}$$

where

$$\Phi(x, y) = \frac{1}{2}\int\limits_{x+y}^{2a-|x-y|} Q\left(\frac{s+x-y}{2}, \frac{s-x+y}{2}\right)\, ds, \tag{2.5.12}$$

$$Q(x, y) = \frac{1}{R}\left(\mathcal{L}_1(-y)\mathcal{L}_2(x) - \mathcal{L}_2(-y)\mathcal{L}_1(x)\right). \tag{2.5.13}$$

Taking into account the transition from the segment $[0, \omega]$ to the segment $[-a, a]$, we deduce from Theorem 2.2.1 the following assertion.

Proposition 2.5.1. If the operator S is regular and if $\varphi(x) \in W_p^{(2)}$ $(1 \leq p \leq 2)$, then $T\varphi \in L^p(-a, a)$ and the equality $ST\varphi = \varphi$ is true.

3. In order to estimate the kernel $\Phi(x, y)$ we introduce the function

$$
h(x) = \begin{cases} \displaystyle\int_{v-a}^{a} |Q(u, u - v)|\, du, & 0 \le v \le 2a, \\[4mm] \displaystyle\int_{-v-a}^{a} |Q(u + v, u)|\, du, & -2a \le v \le 0, \end{cases} \qquad v = x - y.
$$

From (2.5.12), (2.5.13) we have

$$
\left|\Phi(x, y)\right| \le h(x - y), \qquad \int_{-2a}^{2a} h(v)\, dv < \infty. \tag{2.5.14}
$$

It means that the operator

$$
\mathcal{B}\varphi = \int_{-a}^{a} \Phi(x, y)\varphi(y)\, dy \tag{2.5.15}
$$

is bounded in $L(-a, a)$.

Theorem 2.5.2. *If the operator S of the form (2.5.1) is regular and $\varphi(x) \in L(-a,a)$, then equation (2.5.4) has one and only one solution $f = \mathcal{B}\varphi$ which satisfies condition (2.5.3).*

Proof. The equation (2.5.4) is equivalent to the relation

$$
-Sf' = \int_{-a}^{x} \varphi(y)\, dy + C,
$$

where C is a constant. Using (2.5.1) and (2.5.11) we obtain:

$$
f'(x) = \frac{d}{dx}(\mathcal{B}\varphi) - \frac{1}{R} \int_{-a}^{a} \left(\int_{-a}^{y} \varphi(u)\, du + C \right) \mathcal{L}_1(-y)\, dy \mathcal{L}_1(x). \tag{2.5.16}
$$

Since $\Phi(\pm a, y) = 0$, from (2.5.16) and the requirement $f(\pm a) = 0$, we derive $f(x) = \mathcal{B}\varphi$. Hence, the assertion of the theorem follows. $\square$

Chapter 3

Examples and Applications

In this chapter, we consider a number of concrete examples and find the corresponding functions $\mathcal{L}_k(x)$ satisfying the equation

$$S\mathcal{L}_k = x^{k-1}, \quad k = 0, 1. \tag{3.0.1}$$

Then, using the results of Chapters 1 and 2, we present a method for constructing the solution of the equation

$$Sf = \varphi \tag{3.0.2}$$

with an arbitrary right-hand side $\varphi(x)$. We note that the equations considered in this chapter play an essential role in many applied and theoretical problems and have been studied by many authors (see [12, 17, 38] and references therein). Our approach, which was developed in the first and second chapters, enables us to recover and use explicit solutions in some important cases, where only approximate solutions were used before.

In particular, we deal here with various applications to the contact theory of elasticity, hydrodynamics and the theory of radiation transfer. Our procedure to recover explicitly the solution $f(x)$ of the equation (2.0.2) from the known solutions $\mathcal{L}_1$ and $\mathcal{L}_2$ is actively used in these applications.

The chapter also contains several new theoretical results (see Section 3.6).

3.1 Integral equations with kernel of power type

1. In contact problems of elasticity theory, the following equation plays an important role [1, 102]:

$$S_\alpha f = \int\limits_0^\omega \frac{1 - \beta \, \text{sgn}\,(x - t)}{|x - t|^{\alpha - 1}} f(t)\, \mathrm{d}t = \varphi(x), \tag{3.1.1}$$

where

$$-1 < \beta < 1, \quad 0 < \alpha < 2, \quad \alpha \neq 1. \tag{3.1.2}$$

Equation (3.1.1) is also used in the theory of stable processes [30, 33, 87]. The case $\beta = 0$, $1 < \alpha < 2$ was investigated by T. Carleman [12] by methods from the theory of functions of complex variables. The solution found by Carleman contains contour integrals and is well defined only for analytical right-hand sides. For the same case $\beta = 0$ and $1 < \alpha < 2$, M.G. Kreĭn [21, Chap. IV, §8] found a formula for the solution of equation (3.1.1) without using the theory of functions of complex variables. The case $-1 < \beta < 1$ and $1 < \alpha < 2$ is considered by N.K. Arutunyan in his work [4].

2. We shall show that the functions $\mathcal{L}_1(x)$, $\mathcal{L}_2(x)$ corresponding to equation (3.1.1) have the form

$$\mathcal{L}_1(x) = \mathcal{D}x^{-\rho}(\omega - x)^{\rho - \mu}, \tag{3.1.3}$$

$$\mathcal{L}_2(x) = \frac{\omega(\rho - \mu) + x}{1 - \mu}\,\mathcal{L}_1(x), \tag{3.1.4}$$

where the following notations are used:

$$\mu = 2 - \alpha, \tag{3.1.5}$$

$$\mathcal{D} = \frac{\sin \pi\rho}{\pi(1 - \beta)}, \tag{3.1.6}$$

and the value ρ is defined with the help of the relation

$$\tan \pi\rho = \frac{(1 - \beta)\sin \pi\mu}{(1 + \beta) + (1 - \beta)\cos \pi\mu}, \quad 0 < \rho < 1. \tag{3.1.7}$$

We note that the equality $S_\alpha \mathcal{L}_1 = \mathbb{1}$ was proved earlier (see [4]) in a different way. Formula (3.1.7) can be rewritten in the form

$$\sin \pi\rho = \frac{1 - \beta}{1 + \beta}\,\sin \pi(\mu - \rho). \tag{3.1.8}$$

Hence, it follows that

$$0 < \mu - \rho < 1. \tag{3.1.9}$$

Theorem 3.1.1. *Let the operator S_α be defined by the relations (3.1.1), (3.1.2) and the functions $\mathcal{L}_1(x)$, $\mathcal{L}_2(x)$ be defined by the relations (3.1.3)–(3.1.7). Then the equations*

$$S_\alpha \mathcal{L}_k = x^{k-1}, \quad k = 1, 2 \tag{3.1.10}$$

are valid.

Proof. Making substitutions $x = \omega u$, $t = \omega s$, we write

$$S_\alpha \mathcal{L}_1 = \mathcal{D}\int_0^1 s^{-\rho}(1 - s)^{\rho - \mu}\left(1 - \beta\,\mathrm{sgn}\,(u - s)\right)|u - s|^{\mu - 1}\,ds. \tag{3.1.11}$$

Now we represent (3.1.11) in the form

$$
S_\alpha \mathcal{L}_1 = \mathcal{D} \left(\int_0^u s^{-\rho}(1-s)^{\rho-\mu}(u-s)^{\mu-1}\, ds\,(1-\beta) \right.
$$

$$
\left. + \int_u^1 s^{-\rho}(1-s)^{\rho-\mu}(s-u)^{\mu-1}\, ds\,(1+\beta) \right). \tag{3.1.12}
$$

Putting $s = zu$ in the right-hand side of (3.1.12) in the first integral and in the second one $s = 1 - (1-u)z$, we have

$$
S_\alpha \mathcal{L}_1 = \mathcal{D} \left(u^{\mu-\rho} \int_0^1 z^{-\rho}(1-zu)^{\rho-\mu}(1-z)^{\mu-1}\, dz\,(1-\beta) \right.
$$

$$
\left. + (1-u)^\rho \int_0^1 (1-z)^{\mu-1} z^{\rho-\mu}\left(1-(1-u)z\right)^{-\rho}\, dz\,(1+\beta) \right). \tag{3.1.13}
$$

Now we express the integrals standing on the right-hand side of (3.1.13) using hypergeometric function [7, Chap. 2]

$$
F(a,b,c,z) = \frac{\Gamma(c)}{\Gamma(b)\Gamma(c-b)} \int_0^1 \frac{t^{b-1}(1-t)^{c-b-1}}{(1-tz)^a}\, dt \tag{3.1.14}
$$

where $\Gamma(z)$ is Euler's gamma-function. Then formula (3.1.13) gets the form

$$
S_\alpha \mathcal{L}_1 = \mathcal{D}\Gamma(\mu) \left(u^{\mu-\rho}\frac{\Gamma(1-\rho)}{\Gamma(1-\rho+\mu)} F(\mu-\rho,1-\rho,1-\rho+\mu,u)\,(1-\beta) \right.
$$

$$
\left. + (1-u)^\rho \frac{\Gamma(1+\rho-\mu)}{\Gamma(1+\rho)} F(\rho,1+\rho-\mu,1+\rho,1-u)\,(1+\beta) \right). \tag{3.1.15}
$$

The incomplete beta-function

$$
B_u(p,q) = \int_0^u s^{p-1}(1-s)^{q-1}\, ds \tag{3.1.16}
$$

is connected with the hypergeometric function by the following relation (see [7, Chap. 2])

$$
F(p,1-q,p+1,u) = pu^{-p}B_u(p,q). \tag{3.1.17}
$$

Taking into account formulas (3.1.15)–(3.1.17) we deduce

$$
S\mathcal{L}_1 = \mathcal{D}\left(\frac{\Gamma(1-\rho)\Gamma(\mu)}{\Gamma(\mu-\rho)} (1-\beta) \int_0^u t^{\mu-\rho-1}(1-t)^{\rho-1}\, dt \right.
$$

$$
\left. + \frac{\Gamma(1+\rho-\mu)\,\Gamma(\mu)}{\Gamma(\rho)} (1+\beta) \int_0^{1-u} t^{\rho-1}(1-t)^{\mu-\rho-1}\, dt \right). \qquad (3.1.18)
$$

Using the property of the gamma-function

$$
\Gamma(z)\Gamma(1-z) = \frac{\pi}{\sin \pi z}, \qquad (3.1.19)
$$

and the relation (3.1.8), we obtain the equality

$$
\frac{\Gamma(1-\rho)\,(1-\beta)}{\Gamma(\mu-\rho)} = \frac{\Gamma(1+\rho-\mu)\,(1+\beta)}{\Gamma(\rho)}. \qquad (3.1.20)
$$

Since

$$
\int_0^{1-u} t^{\rho-1}(1-t)^{\mu-\rho-1}\, dt = \int_u^1 t^{\mu-\rho-1}(1-t)^{\rho-1}\, dt,
$$

by virtue of (3.1.20) formula (3.1.18) takes the form

$$
S\mathcal{L}_1 = \mathcal{D}\frac{\Gamma(1-\rho)\Gamma(\mu)}{\Gamma(\mu-\rho)} (1-\beta) \int_0^1 t^{\mu-\rho-1}(1-t)^{\rho-1}\, dt. \qquad (3.1.21)
$$

From the equality

$$
\int_0^1 t^{\mu-\rho-1}(1-t)^{\rho-1}\, dt = \frac{\Gamma(\mu-\rho)\Gamma(\rho)}{\Gamma(\mu)} \qquad (3.1.22)
$$

and the formulas (3.1.6), (3.1.19), (3.1.21) it follows that $S\mathcal{L}_1 = 1$. Thus, the equality (3.1.10) when $k = 1$ is proved. In order to prove the equality (3.1.10) when $k = 2$ we consider the expression

$$
S_\alpha(x\mathcal{L}_1) = \mathcal{D}\left(\int_0^x t^{1-\rho}(\omega-t)^{\rho-\mu}(x-t)^{\mu-1}\, dt\,(1-\beta) \right.
$$

$$
\left. + \int_x^\omega t^{1-\rho}(\omega-t)^{\rho-\mu}(t-x)^{\mu-1}\, dt\,(1+\beta) \right). \qquad (3.1.23)
$$

Putting $x = \omega u$, $t = \omega u z$ in the right-hand side of (3.1.23) in the first integral and in the second one $x = \omega u$, $t = \omega[1 - (1 - u)z]$ we have

$$S_\alpha(x\mathcal{L}_1) = \mathcal{D}\omega \left(u^{\mu - \rho + 1} \int_0^1 z^{1-\rho}(1 - zu)^{\rho - \mu}(1 - z)^{\mu - 1}\, dz\, (1 - \beta) \right. \tag{3.1.24}$$

$$\left. + (1 - u)^\rho \int_0^1 (1 - z)^{\mu - 1} z^{\rho - \mu} \left(1 - (1 - u)z\right)^{1-\rho}\, dz\, (1 + \beta) \right).$$

Expressing the integrals, standing in the right-hand side of (3.1.24) by the hypergeometric function, we obtain

$$S_\alpha(x\mathcal{L}_1) = \mathcal{D}\omega\Gamma(\mu) \left(u^{\mu - \rho + 1} \frac{\Gamma(2 - \rho)}{\Gamma(2 - \rho + \mu)} F\left(\mu - \rho, 2 - \rho, 2 - \rho + \mu, u\right) (1 - \beta) \right.$$

$$\left. + (1 - u)^\rho \frac{\Gamma(1 + \rho - \mu)}{\Gamma(1 + \rho)} F\left(\rho - 1, 1 + \rho - \mu, 1 + \rho, 1 - u\right) (1 + \beta) \right). \tag{3.1.25}$$

Now we use Gauss' relation [7, Ch.2]

$$F(a, b, c, z) = -\frac{a}{c - a - 1} F(a + 1, b, c, z) + \frac{c - 1}{c - a - 1} F(a, b, c - 1, z). \tag{3.1.26}$$

From formulas (3.1.25) and (3.1.26) we deduce that

$$S_\alpha(x\mathcal{L}_1) = \mathcal{D}\omega\Gamma(\mu)$$

$$\times \left(u^{\mu - \rho + 1} \frac{\Gamma(2 - \rho)}{\Gamma(2 - \rho + \mu)} \left((\rho - \mu) F(\mu - \rho + 1, 2 - \rho, 2 - \rho + \mu, u) \right.\right.$$

$$\left. + (1 - \rho + \mu) F(\mu - \rho, 2 - \rho, 1 - \rho + \mu, u) \right) (1 - \beta)$$

$$+ (1 - u)^\rho \frac{\Gamma(1 + \rho - \mu)}{\Gamma(1 + \rho)} \left((1 - \rho) F(\rho, 1 + \rho - \mu, 1 + \rho, 1 - u) \right.$$

$$\left.\left. + \rho F(\rho - 1, 1 + \rho - \mu, \rho, 1 - u) \right) (1 + \beta) \right). \tag{3.1.27}$$

From (3.1.20) we have the equation

$$\frac{\Gamma(1 + \rho - \mu)(1 + \beta)}{\Gamma(1 + \rho)} = \frac{\Gamma(2 - \rho)(1 - \beta)}{\Gamma(\mu - \rho)\rho(1 - \rho)}. \tag{3.1.28}$$

Thus, the equality (3.1.27) can be written in the form

$$S_\alpha(x\mathcal{L}_1) = \mathcal{D}\omega\Gamma(\mu)\frac{\Gamma(2-\rho)(1-\beta)}{\Gamma(\mu-\rho)}$$

$$\times\left[u^{\mu-\rho+1}\left(-\frac{1}{\mu-\rho+1}F\left(\mu-\rho+1,2-\rho,2-\rho+\mu,u\right)\right.\right.$$

$$\left.+\frac{1}{\mu-\rho}F\left(\mu-\rho,2-\rho,1-\rho+\mu,u\right)\right)$$

$$+(1-u)^\rho\left(\frac{1}{\rho}F\left(\rho,1+\rho-\mu,1+\rho,1-u\right)\right.$$

$$\left.\left.+\frac{1}{1-\rho}F\left(\rho-1,1+\rho-\mu,\rho,1-u\right)\right)\right]. \qquad (3.1.29)$$

We denote the expression standing in the square brackets on the right-hand side of (3.1.29) by $\Phi(\rho,\mu,u)$. Using (3.1.16), (3.1.17) we have

$$\Phi(\rho,\mu,u) = \int_0^1 t^{\mu-\rho-1}(1-t)^{\rho-2}(u-t)\,\mathrm{d}t$$

$$= -\frac{\Gamma(\mu-\rho+1)\Gamma(\rho-1)}{\Gamma(\mu)} + u\frac{\Gamma(\mu-\rho)\Gamma(\rho-1)}{\Gamma(\mu-1)}. \qquad (3.1.30)$$

We remark that formulas (3.1.16), (3.1.17) are well defined under the condition $p > 0$, $q < 0$. It means that (3.1.30) is deduced under the condition $\rho > 1$, $\mu > \rho$. However, in view of the analyticity with respect to ρ, relation (3.1.30) remains valid in the domain $0 < \rho < 1$, $\mu > \rho$ as well. According to (3.1.6), (3.1.19) the equality

$$\mathcal{D}\,\Gamma(2-\rho)\,\Gamma(\rho-1)\,(1-\beta) = -1 \qquad (3.1.31)$$

is valid. It follows from (3.1.29)–(3.1.31) that

$$S_\alpha(x\mathcal{L}_1) = -x(\mu-1) + \omega(\mu-\rho). \qquad (3.1.32)$$

From (3.1.32) the validity of the equality (3.1.10) when $k = 2$ follows directly. This proves the theorem. $\qquad\square$

From (3.1.3) and (3.1.22) we deduce that

$$R = \int_0^\omega \mathcal{L}_1(x)\,\mathrm{d}x = \mathcal{D}\omega^{1-\mu}\frac{\Gamma(1-\rho)\Gamma(1+\rho-\mu)}{\Gamma(2-\mu)}. \qquad (3.1.33)$$

From equalities (3.1.3), (3.1.4) and formula (2.2.26) we have

$$Q_1(x,t) = \frac{\mathcal{D}^2}{R(1-\mu)^2}\,x^{-\rho}(\omega-x)^{\rho-\mu}(\omega-t)^{-\rho}t^{\rho-\mu}(x+t-\omega). \qquad (3.1.34)$$

Substituting $s = u + \omega$ into (2.2.30), from (3.1.34) we obtain

$$\Phi_1(x,t) = \frac{\mathcal{D}^2}{2R(1-\mu)^2} \tag{3.1.35}$$

$$\times \int\limits_{x+t-\omega}^{\omega-|x-t|} \left(\frac{(\omega+x-t)^2 - u^2}{4}\right)^{-\rho} \left(\frac{(\omega-x+t)^2 - u^2}{4}\right)^{\rho-\mu} u\, du.$$

With the help of the formulas (2.2.31) and (3.1.35) we construct the operator T_α (let us remember that $N_2(x) = \mathcal{L}_1(x)$). From the relations (3.1.3), (3.1.4) and (2.2.32) it follows that $\gamma = 0$. Then the following equality is valid:

$$S_\alpha T_\alpha \varphi = \varphi, \qquad \varphi \in W_p^{(2)}. \tag{3.1.36}$$

Recall that under the condition $1 < \alpha < 2$, the solutions $\mathcal{L}_1$ and $\mathcal{L}_2$ for equation (3.1.1) were found in the work [59] by S. Pozin.

3. In the important subcase

$$\beta = 0, \tag{3.1.37}$$

our formulas take a simpler form. From (3.1.6) and (3.1.7) it follows that

$$\rho = \frac{\mu}{2}, \quad \mathcal{D} = \frac{\sin(\pi\mu/2)}{\pi}. \tag{3.1.38}$$

In this case the functions $\mathcal{L}_1(x)$ and $\mathcal{L}_2(x)$ have the form

$$\mathcal{L}_1(x) = \frac{\sin(\pi\mu/2)}{\pi}(x(\omega - x))^{-\mu/2}, \quad \mathcal{L}_2(x) = \frac{x - \mu a}{1 - \mu}\mathcal{L}_1(x), \tag{3.1.39}$$

where $a = \omega/2$. From (3.1.33) we obtain

$$R = \int\limits_0^\omega \mathcal{L}_1(x)\, dx = \frac{\sin(\pi\mu/2)}{\pi}\omega^{1-\mu}\frac{\Gamma^2(1-\mu/2)}{\Gamma(2-\mu)}. \tag{3.1.40}$$

3.2 Integral equations with logarithmic type kernels

1. The principle equation of the planar contact theory of elasticity has the form [102]

$$S_b f = \int\limits_0^\omega \ln\left(\frac{b}{2|x-t|}\right) f(t)\, dt = \varphi(x), \quad b \neq \frac{\omega}{2}, \quad b > 0. \tag{3.2.1}$$

T. Carleman [12] studied (3.2.1) using methods of the theory of functions of a complex variable. In later works [17, 38], equation (3.2.1) was solved by some other methods.

2. Let us introduce the notations

$$\mathcal{L}_1(x) = \frac{R}{\pi\sqrt{x(\omega - x)}}, \quad \mathcal{L}_2(x) = a\mathcal{L}_1(x) + \frac{1}{\pi}\frac{x - a}{\sqrt{x(\omega - x)}}, \tag{3.2.2}$$

where

$$a = \frac{\omega}{2}, \quad R = \int_0^\omega \mathcal{L}_1(t)\,dt = \left(\ln\frac{b}{a}\right)^{-1}. \tag{3.2.3}$$

The equality $S_b\mathcal{L}_1 = 1$ was proved earlier (see [12, 38]). From Theorem 3.1.1 when $\mu \to 1$ we obtain the following assertion.

Theorem 3.2.1. *Let the operator S_b have the form* (3.2.1) *and the function $\mathcal{L}_1(x)$, $\mathcal{L}_2(x)$ be defined by* (3.2.2), (3.2.3). *Then the equalities*

$$S_b\mathcal{L}_k = x^{k-1}, \quad k = 1, 2 \tag{3.2.4}$$

are valid.

Proof. By (3.1.10), (3.1.39), (3.1.40) we obtain

$$\frac{1}{\pi}\int_0^\omega \left(\sin\frac{\pi\mu}{2}\right)(t(\omega - t))^{-\mu/2}\left(|x - t|^{\mu-1} - 1\right)dt$$

$$= 1 - \frac{\sin(\pi\mu/2)}{\pi\,\Gamma(2 - \mu)}\,\omega^{1-\mu}\,\Gamma^2\left(1 - \frac{\mu}{2}\right). \tag{3.2.5}$$

Recall the well-known (see, e.g., [7, Chap. 1]) equalities

$$\Gamma(1) = 1, \quad \Gamma(1/2) = \sqrt{\pi}, \quad \Gamma'(1) = C, \tag{3.2.6}$$

$$\Gamma'(1/2) = \sqrt{\pi}(-C - \ln 4), \tag{3.2.7}$$

where C is Euler's constant. Dividing (3.2.5) by $(\mu - 1)$ when $\mu \to 1$ we obtain

$$\frac{1}{\pi}\int_0^\omega (t(\omega - t))^{-\frac{1}{2}}\ln|x - t|\,dt = -\frac{1}{\pi}\left(\frac{d}{d\mu}\frac{\Gamma^2(1 - \mu/2)}{\Gamma(2 - \mu)}\right)\bigg|_{\mu=1} + \ln\omega. \tag{3.2.8}$$

From (3.2.6)–(3.2.8) we deduce the relation

$$\frac{1}{\pi}\int_0^\omega (t(\omega - t))^{-\frac{1}{2}}\ln|x - t|\,dt = \ln\frac{a}{2}. \tag{3.2.9}$$

Hence, the formula (3.2.4) is valid when $k = 1$. For proving the formula (3.2.4) in the case $k = 2$, we shall use the relations (3.1.10) and (3.1.39):

$$\int_0^\omega \frac{\sin(\pi\mu/2)}{\pi} \frac{t - \mu a}{1 - \mu} \left(t(\omega - t)\right)^{-\mu/2} \left(|x - t|^{\mu-1} - 1\right) dt$$

$$= x + \frac{\mu a \sin(\pi\mu/2)}{\pi(1 - \mu)\Gamma(2 - \mu)} \omega^{1-\mu} \Gamma^2\left(1 - \frac{\mu}{2}\right)$$

$$- \frac{\sin(\pi\mu/2)}{\pi(1 - \mu)} \omega^{2-\mu} \frac{\Gamma(2 - \mu/2)\,\Gamma(1 - \mu/2)}{\Gamma(3 - \mu)}$$

$$= x + \frac{\sin(\pi\mu/2)}{\pi\Gamma(2 - \mu)} \omega^{2-\mu} \Gamma^2\left(1 - \frac{\mu}{2}\right)\left(-\frac{1}{2}\right). \tag{3.2.10}$$

Here the well-known equalities (see [7, Chap. 1])

$$\int_0^1 t^\alpha (1 - t)^\beta \, dt = \frac{\Gamma(\alpha + 1)\Gamma(\beta + 1)}{\Gamma(\alpha + \beta + 2)}, \quad \Gamma(z + 1) = z\,\Gamma(z)$$

are used. From (3.2.10) when $\mu \to 1$ we have

$$-\frac{1}{\pi} \int_0^\omega (t - a)\left(t(\omega - t)\right)^{-1/2} \ln|x - t| \, dt = x - a. \tag{3.2.11}$$

This proves the theorem. $\qquad\qquad\qquad\qquad\qquad\qquad\qquad\qquad\square$

According to (2.2.26) the equality

$$Q_1(x, t) = \frac{1}{\pi^2} \frac{x + t - \omega}{\sqrt{xt(\omega - x)(\omega - t)}} \tag{3.2.12}$$

is valid. Then, taking into account (2.2.30), we see that

$$\Phi_1(x, t) = -\frac{1}{\pi^2} \ln \frac{\omega|x - t|}{\left(\sqrt{x(\omega - t)} + \sqrt{t(\omega - x)}\right)^2}. \tag{3.2.13}$$

It follows from (2.2.32) and (3.2.2), (3.2.3), (3.2.13) that the operator T has the form

$$T\varphi = \frac{1}{\pi^2} \frac{d}{dx} \int_0^\omega \varphi'(t) \ln \frac{\omega|x - t|}{\left(\sqrt{x(\omega - t)} + \sqrt{t(\omega - x)}\right)^2} \, dt$$

$$+ \frac{R}{\pi^2} \int_0^\omega \varphi(t) \frac{1}{\sqrt{t(\omega - x)}} \, dt \, \frac{1}{\sqrt{x(\omega - t)}}. \tag{3.2.14}$$

Theorem 3.2.2. *If* $\varphi'(x) \in L^p(0,\omega)$ $(1 < p \leq 2)$, *then equation* (3.2.1) *has one and only one solution*

$$f(x) = T\varphi \tag{3.2.15}$$

in $L^r(0,\omega)$ $(1 \leq r < p)$.

Proof. By M. Riesz' theorem (see [104, Chap. 5]) the operator

$$Vf = \frac{\mathrm{d}}{\mathrm{d}x} \int_0^\omega f(t) \ln|x - t| \, \mathrm{d}t \tag{3.2.16}$$

is bounded in $L^p(0,\omega)$ $(1 < p)$. We consider the operator

$$\widetilde{T}\varphi = \frac{\mathrm{d}}{\mathrm{d}x} \int_0^\omega \varphi(t) \ln\left(\sqrt{(\omega - t)x} + \sqrt{t(\omega - x)}\right)^2 \mathrm{d}t$$

$$= \frac{1}{\sqrt{x(\omega - x)}} \int_0^\omega \varphi(t) \frac{\sqrt{(\omega - t)(\omega - x)} - \sqrt{xt}}{\sqrt{(\omega - t)x} + \sqrt{t(\omega - x)}} \, \mathrm{d}t. \tag{3.2.17}$$

From the last equation we obtain

$$\left|\left(\widetilde{T}\varphi\right)(x)\right| \leq \frac{1}{\sqrt{x(\omega - x)}} \|\varphi\|_p \left(\int_0^\omega \left|\frac{1 + \xi z}{\xi + z}\right|^q \mathrm{d}t\right)^{1/q}, \tag{3.2.18}$$

where

$$\xi = \sqrt{\frac{x}{\omega - x}}, \quad z = \sqrt{\frac{t}{\omega - t}}, \quad \frac{1}{p} + \frac{1}{q} = 1.$$

In order to estimate the integral on the right-hand side of (3.2.18) we use the inequalities

$$\frac{1 + \xi z}{\xi + z} \leq \max\left\{z, z^{-1}\right\}, \quad \frac{1 + \xi z}{\xi + z} \leq \max\left\{\xi, \xi^{-1}\right\}. \tag{3.2.19}$$

Let $x \leq \dfrac{\omega}{2}$. Then $\xi \leq 1$ and

$$\int_0^\omega \left|\frac{1 + \xi z}{\xi + z}\right|^q \mathrm{d}t \leq \int_0^x \xi^{-q} \, \mathrm{d}t + \int_x^{\frac{\omega}{2}} z^{-q} \, \mathrm{d}t + \int_{\frac{\omega}{2}}^{\omega - x} z^q \, \mathrm{d}t + \int_{\omega - x}^\omega \xi^q \, \mathrm{d}t \leq C x^{1 - q/2}. \tag{3.2.20}$$

Similarly, for $x \geq \dfrac{\omega}{2}$ we obtain

$$\int_0^\omega \left|\frac{1 + \xi z}{\xi + z}\right|^q \mathrm{d}t \leq C(\omega - x)^{1 - q/2}. \tag{3.2.21}$$

Comparing (3.2.18), (3.2.20) and (3.2.21) we see that

$$\left|\left(\widetilde{T}\varphi\right)(x)\right| \leq C\left(x(\omega - x)\right)^{-1/p}\|\varphi\|_p. \tag{3.2.22}$$

Taking an arbitrary $r \in (1, p)$ we have

$$T\varphi \in L^r(0, \omega), \qquad 1 \leq r < p. \tag{3.2.23}$$

From Theorem 2.2.5 and (3.2.1), (3.2.2) it follows that $\gamma = 0$. Then (2.2.7) holds for $\varphi''(x) \in L^p(0, \omega)$. In view of (3.2.23) it holds also for $\varphi'(x) \in L^p(0, \omega)$, that is

$$ST\varphi = \varphi, \qquad \varphi' \in L^p(0, \omega). \tag{3.2.24}$$

Hence, the function $f(x) = T\varphi$, which lies in $L^r(0, \omega)$, is a solution of (3.2.1).

Using (2.2.25) and (3.2.1), (3.2.2) we can verify that the conditions of Theorem 2.2.5 are satisfied for any p $(1 \leq p < 2)$. Consequently, (3.2.1) has no more than one solution in $L(0, \omega)$. This proves the theorem. $\square$

This theorem improves a result of Tricomi [105] in which it is required that $r < \dfrac{4}{3} < p$.

Remark 3.2.3. If we take into account that

$$\frac{1}{x - t} - \frac{1}{\sqrt{x(\omega - x)}} \frac{\sqrt{(\omega - t)(\omega - x)} - \sqrt{xt}}{\sqrt{(\omega - t)x} + \sqrt{t(\omega - x)}} = \frac{\sqrt{t(\omega - t)}}{\sqrt{x(\omega - x)}(x - t)} \tag{3.2.25}$$

then we obtain from (3.2.14), (3.2.17) the representation of T obtained by Carleman [12]

$$T\varphi = \frac{1}{\pi^2} \int_0^\omega \varphi'(t) \frac{\sqrt{t(\omega - t)}}{\sqrt{x(\omega - x)}(x - t)} \, dt + \frac{1}{\pi^2 \ln \frac{b}{a} \sqrt{x(\omega - x)}} \int_0^\omega \frac{\varphi(t) \, dt}{\sqrt{t(\omega - t)}}. \tag{3.2.26}$$

3. We consider the following triangular operator

$$Sf = \int_0^x f(t) \ln(x - t) \, dt, \qquad f(x) \in L(0, \omega) \tag{3.2.27}$$

and introduce the function

$$E(x) = \int_0^\infty e^{-C\xi} x^{\xi - 1} \frac{1}{\Gamma(\xi)} \, d\xi, \tag{3.2.28}$$

where C is Euler's constant. It follows from (3.2.28) that for any m

$$E(x) \leq m \int_0^1 x^{\xi-1} \xi \, d\xi \leq \frac{2m}{x \ln^2 x}, \tag{3.2.29}$$

that is, $E(x) \in L(0, \omega)$.

Let us calculate the function

$$SE(x) = \int_0^\infty e^{-C\xi} \frac{1}{\Gamma(\xi)} \int_0^x t^{\xi-1} \ln(x - t) \, dt \, d\xi \tag{3.2.30}$$

by using the formula

$$\left(\frac{\partial}{\partial \alpha} \int_0^x (x-t)^{\alpha-1} t^{\xi-1} \, dt \right) \Bigg|_{\alpha=1} = \int_0^x t^{\xi-1} \ln(x-t) \, dt. \tag{3.2.31}$$

Taking into account the relation

$$\int_0^x (x-t)^{\alpha-1} t^{\xi-1} \, dt = x^{\alpha+\xi-1} \frac{\Gamma(\alpha)\Gamma(\xi)}{\Gamma(\alpha+\xi)} \tag{3.2.32}$$

we deduce from (3.2.31) that

$$\int_0^x t^{\xi-1} \ln(x-t) \, dt = x^\xi \left((\ln x) \frac{1}{\Gamma(1+\xi)} + \frac{\Gamma'(1)}{\Gamma(1+\xi)} - \frac{\Gamma'(1+\xi)}{\Gamma^2(1+\xi)} \right) \Gamma(\xi). \tag{3.2.33}$$

Hence, $\Gamma'(1) = -C$. Then by (3.2.33) we obtain

$$SE(x) = \int_0^\infty \frac{d}{d\xi} \left(e^{-C\xi} x^\xi \frac{1}{\Gamma(1+\xi)} \right) d\xi = -1.$$

Thus, it is proved that

$$N_2(x) = -E(x). \tag{3.2.34}$$

The formula (3.2.34) can be also deduced with the help of the Laplace transformation.

According to Theorem 2.2.5 the equality $\gamma = 0$ is valid. Let us remark that for triangular operators S of the form (3.2.1) the relation $N_1(x) = 1$ holds. Then from (2.2.12) and (3.2.34) we deduce the formula $Q(x, t) = -E(\omega - t)$. In view of (2.2.13) we have

$$\Phi(x, t) = -\frac{1}{2} \int_{x+t}^{2\omega - |x-t|} E\left(\omega - \frac{s - x + t}{2} \right) ds.$$

Taking into account (2.2.15) we obtain

$$T\varphi = -\varphi(0)E(x) - \int\limits_0^x \varphi'(t)E(x-t)\,dt. \qquad (3.2.35)$$

If $\varphi(x) \in W_1^{(1)}$, formula (3.2.35) implies that $T\varphi \in L(0,\omega)$. This enables us to modify and improve Theorem 2.2.1 for the present case.

Theorem 3.2.4. *For the operators S and T defined by (3.2.1) and (3.2.35), the following equality is valid:*

$$ST\varphi = \varphi, \qquad \varphi \in W_1^{(1)}(0,\omega).$$

3.3 Regularization

In order to analyze equations with a difference kernel (similar to those studied in Sections 3.1 and 3.2) we use the method of regularization. By treating the regularized equation we can prove (in a number of cases) the existence of the solutions N_1 and N_2, and sometimes calculate these solutions with the required in the applications accuracy. However, after this, it makes sense to return to the original equation in order to utilize fully the specific structure of the kernel.

1. We illustrate the method of regularization on equations of the form

$$\mathcal{H}f = \int\limits_0^\omega \left(\ln \frac{b}{2|x-t|} + h(x-t) \right) f(t)\,dt = \varphi(x) \qquad (3.3.1)$$

where

$$f(x) \in L^p(0,\omega) \quad (1 < p < 2), \quad b > 0, \quad b \neq \frac{\omega}{2}, \qquad (3.3.2)$$

and we assume that for some c, the inequality

$$\int\limits_0^\omega |h'(u)|^q (\omega - |u|)\,du \le c, \quad \frac{1}{q} + \frac{1}{p} = 1 \qquad (3.3.3)$$

holds. It was proved in Theorem 3.2.2 that

$$ST\varphi = \varphi, \quad \varphi' \in L^p(0,\omega), \quad (1 < p < 2), \qquad (3.3.4)$$

where the operators S and T are defined by (3.2.1) and (3.2.14). If $f \in L^p(0,\omega)$ $(1 < p < 2)$, then the function

$$\frac{d}{dx}Sf = \int\limits_0^\omega \frac{f(t)}{x-t}\,dt$$

belongs to $L^p(0, \omega)$, that is, Sf belongs to the domain of definition of T. It follows from Theorem 3.2.2 that the function

$$g(x) = TSf \tag{3.3.5}$$

belongs to $L^r(0, \omega)$ $(1 < r < p)$. Applying S to both sides of (3.3.5) and using (3.3.4), we obtain

$$Sg = Sf. \tag{3.3.6}$$

Since (3.2.1) has at most one solution in $L^p(0, \omega)$, it follows from (3.3.5) and (3.3.6) that

$$TSf = f. \tag{3.3.7}$$

From (3.3.1) and (3.3.7), using traditional methods we deduce the following result.

Theorem 3.3.1. *Suppose that (3.3.3) is satisfied. Then*

$$T\mathcal{H} = I + B, \tag{3.3.8}$$

where B is compact in $L^p(0, \omega)$.

Corollary 3.3.2. *Suppose that (3.3.3) holds and*

$$\varphi'(x) \in L^{p_1}(0, \omega) \qquad (1 < p < p_1 \le 2).$$

I. *The regularized equation*

$$(I + B)f = Tf \tag{3.3.9}$$

   is equivalent to the original equation, that is, any solution of one of them is also a solution of the another one.

II. *If the equation*

$$\mathcal{H}f = 0 \tag{3.3.10}$$

   has only the trivial solution in $L^p(0, \omega)$, then (3.3.1) has one and only one solution in $L^p(0, \omega)$.

Corollary 3.3.3. *Suppose that (3.3.3) holds and that (3.3.10) has only the trivial solution in $L^p(0, \omega)$. Then there are functions $N_1(x)$ and $N_2(x)$ corresponding to $\mathcal{H}$ in $L^p(0, \omega)$.*

3.4 Fractional integrals of purely imaginary order

1. Let us consider the operator

$$\mathcal{J}^{i\alpha} f = \frac{1}{\Gamma(i\alpha + 1)} \frac{\mathrm{d}}{\mathrm{d}x} \int_0^x (x - t)^{i\alpha} f(t) \, \mathrm{d}t, \quad 0 \le x \le \omega, \tag{3.4.1}$$

where $\alpha = \overline{\alpha}$ and $\Gamma(z)$ is Euler's gamma function. H. Kober [35] proved that $\mathcal{J}^{i\alpha}$ is a bounded operator in $L^2(0, \omega)$. This fact also follows from the general

Theorem 1.5.2 on the boundeness of the operator S. Indeed, in the case (3.4.1) the function $\widetilde{S}(\lambda, \alpha)$ has the form

$$\widetilde{S}(\lambda, \alpha) = \frac{1}{\Gamma(i\alpha + 1)} \int_0^\omega e^{i\lambda t} t^{i\alpha} \, dt. \tag{3.4.2}$$

According to Cauchy theorem we have

$$\int_\gamma e^{i\lambda t} t^{i\alpha} \, dt = 0, \tag{3.4.3}$$

where the curve γ is defined by Figure 3.1.

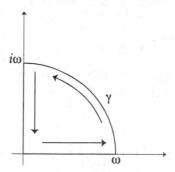

Figure 3.1: Contour of integration.

From (3.4.2) and (3.4.3) we obtain the equality

$$\widetilde{S}(\lambda, \alpha) = \frac{1}{\Gamma(i\alpha + 1)} \tag{3.4.4}$$

$$\times \left((i)^{i\alpha+1} \int_0^\omega e^{-\lambda\nu} \nu^{i\alpha} \, d\nu - i\omega^{i\alpha+1} \int_0^{\pi/2} \exp e^{i\lambda\omega \exp(i\varphi)} e^{i\varphi(i\alpha+1)} \, d\varphi \right).$$

Let us consider the integral

$$\int_0^\omega e^{-\lambda\nu} \nu^{i\alpha} \, d\nu = \left(\int_0^\infty e^{-z} z^{i\alpha} \, dz - \int_{\omega\lambda}^\infty e^{-z} z^{i\alpha} \, dz \right) \lambda^{-i\alpha-1}. \tag{3.4.5}$$

From (3.4.5), when $\lambda \to +\infty$, we deduce

$$\int_0^\omega e^{-\lambda\nu} \nu^{i\alpha} \, d\nu = \Gamma(i\alpha + 1)\lambda^{-i\alpha-1} + O\left(\frac{1}{\lambda^2}\right), \qquad \lambda \to +\infty. \tag{3.4.6}$$

We now estimate the integral

$$\int_0^{\pi/2} e^{i\lambda\omega\,\exp(i\varphi)}e^{i\varphi(i\alpha+1)}i\,d\varphi = \frac{1}{i\lambda\omega}\left(e^{i\lambda\omega\,\exp(i\varphi)}e^{-\alpha\varphi}\right)\Big|_0^{\pi/2} + \frac{\alpha}{i\lambda\omega}\int_0^{\pi/2} e^{i\lambda\omega\,\exp(i\varphi)}e^{-\alpha\varphi}\,d\varphi$$

$$= -\frac{e^{i\lambda\omega}}{i\lambda\omega} + O\left(\frac{1}{\lambda^2}\right). \tag{3.4.7}$$

Thus, from (3.4.4)–(3.4.6) it follows that

$$\widetilde{S}(\lambda,\alpha) = \lambda^{-i\alpha-1}e^{-\alpha\frac{\pi}{2}}i + \frac{1}{\Gamma(i\alpha+1)i\lambda}e^{i\lambda\omega}\omega^{i\alpha} + O\left(\frac{1}{\lambda^2}\right), \quad \lambda \to +\infty. \tag{3.4.8}$$

In the same way we deduce the asymptotic equality

$$\widetilde{S}(\lambda,\alpha) = -|\lambda|^{-i\alpha-1}e^{\alpha\pi/2} + \frac{1}{\Gamma(i\alpha+1)i\lambda}e^{i\lambda\omega}\omega^{i\alpha} + O\left(\frac{1}{\lambda^2}\right), \quad \lambda \to -\infty. \tag{3.4.9}$$

From (3.4.8), (3.4.9) we deduce that $-i\lambda\widetilde{S}(\lambda,\alpha)$ is a bounded function on the axis $(-\infty,\infty)$. By Theorem 1.5.2 the operator $\mathcal{J}^{i\alpha}$ is bounded in $L^2(0,\omega)$.

2. Let us calculate the function

$$\varphi_m(x) = \mathcal{J}^{i\alpha}x^m = \frac{1}{\Gamma(i\alpha+1)}\frac{d}{dx}\int_0^x t^m(x-t)^{i\alpha}\,dt, \quad \operatorname{Re}m > -\frac{1}{2}.$$

Using the equality [7]

$$\int_0^1 h^m(1-h)^{i\alpha}\,dh = \frac{\Gamma(m+1)\Gamma(1+i\alpha)}{\Gamma(m+2+i\alpha)},$$

we obtain

$$\mathcal{J}^{i\alpha}x^m = x^{m+i\alpha}\frac{\Gamma(m+1)}{\Gamma(m+1+i\alpha)}. \tag{3.4.10}$$

From (3.4.10) the important equality

$$\mathcal{J}^{i\alpha}\mathcal{J}^{i\beta} = \mathcal{J}^{i(\alpha+\beta)} \tag{3.4.11}$$

follows. Thus, the operators $\mathcal{J}^{i\alpha}$ have the group property.

3. Let us introduce the operators

$$Af = xf(x) + i\alpha\int_0^x f(t)\,dt, \tag{3.4.12}$$

$$Qf = xf(x), \quad f(x) \in L^2(0,\omega). \tag{3.4.13}$$

It is easy to see that

$$Ax^m = x^{m+1}\frac{m+1+i\alpha}{m+1}, \quad \operatorname{Re} m > -\frac{1}{2}. \tag{3.4.14}$$

Using the relations (3.4.10) and (3.4.12)–(3.4.14) we obtain the equality

$$A = \mathcal{J}^{-i\alpha}Q\mathcal{L}^{i\alpha}. \tag{3.4.15}$$

Thus, the considered operator $\mathcal{J}^{i\alpha}$ transforms the operator A into the diagonal operator Q. This example led to a series of our works on similarity transformations of non-selfadjoint operators (on reduction of operators to a simplest form) [70, 73–76].

Remark 3.4.1. The problem of describing the spectrum of the operator $\mathcal{J}^{i\alpha}$ was formulated in the book [25, Chap. XXIII, §6]. We proved [81] that the spectrum of $\mathcal{J}^{i\alpha}$ coincides with the set

$$e^{-\frac{|\alpha|}{2}\pi} \le |z| \le e^{\frac{|\alpha|}{2}\pi}$$

(see Chapter 8 of this book).

Remark 3.4.2. Let the operator K_α be defined by the formula

$$K_\alpha = (\mathcal{J}^{i\alpha})(\mathcal{J}^{i\alpha})^*. \tag{3.4.16}$$

Then the equalities [74]

$$K_\alpha f = f(x)\cosh(\alpha\pi) + \frac{i}{\pi}\sinh(\alpha\pi)\int_0^\omega \frac{t^{-i\alpha}x^{i\alpha}}{x-t}f(t)\,dt, \tag{3.4.17}$$

$$K_\alpha^{-1}f = f(x)\cosh(\alpha\pi) - \frac{i}{\pi}\sinh(\alpha\pi)\int_0^\omega \frac{(\omega-t)^{-i\alpha}(\omega-x)^{i\alpha}}{x-t}f(t)\,dt \tag{3.4.18}$$

are valid. (The notation $\int_0^\omega$ means the principal value of the Cauchy integral.)

Remark 3.4.3. Formula (3.4.16) gives a non-special triangular factorization of the operator K_α. This case is a good illustration of the general triangular factorizaton approach from [80, 82, 96]. Moreover, this example initiated to a certain degree our work on the general triangular factorizaton [80, 82, 96]. Recall also that the special triangular factorization (i.e., the factorization, where the operator, which is factorized, and both factors may be represented as the sums of the identity operator and compact operators) was introduced and investigated in [21, 38].

3.5   On classes of integral equations which are solvable explicitly

1. Let us introduce the operators

$$S_n f = \int\limits_0^\omega \left(P_m(t-x) \ln|t-x| + Q_n(t-x) \right) f(t)\, dt, \qquad (3.5.1)$$

where $P_m(x)$, $Q_n(x)$ are polynomials and $\deg P_m(x) \le \deg P_n(x) = n$.

Lemma 3.5.1. *The equalities*

$$\fint\limits_0^\omega \frac{dt}{\sqrt{t(\omega-t)}(t-x)} = 0, \quad 0 < x < \omega, \qquad (3.5.2)$$

$$\fint\limits_0^\omega \frac{\ln t\, dt}{\sqrt{t(\omega-t)}(t-x)} = \frac{\pi}{\sqrt{x(\omega-x)}} \left(\frac{\pi}{2} - \arcsin\frac{2x-\omega}{\omega} \right), \quad 0 < x < \omega \quad (3.5.3)$$

are valid.

Proof. Differentiating the equality (3.2.4) when $k = 1$, we obtain the relation (3.5.2). To prove (3.5.3) we introduce the function

$$\Psi(z) = \frac{1}{\sqrt{z(z-\omega)}} \ln(-z),$$

which is defined on the whole plane, except $x \ge 0$. We choose such value of $\ln(-z)$ and $(z(z-\omega))^{-1/2}$ that the relations

$$\Psi_\pm(x) = \lim_{y\to\pm 0} \Psi(x+iy) = (\ln x \pm i\pi)\left(\pm i\sqrt{x(\omega-x)} \right)^{-1} \quad (0 < x < \omega), \quad (3.5.4)$$

$$\Psi_\pm(x) = \lim_{y\to\pm 0} \Psi(x+iy) = (\ln x \pm i\pi)\left(\sqrt{x(x-\omega)} \right)^{-1} \quad (\omega < x < \infty) \qquad (3.5.5)$$

are valid. Since

$$\Psi(x) = \frac{1}{2\pi i} \int\limits_0^\infty \frac{\Psi_+(t) - \Psi_-(t)}{t-z}\, dt,$$

then from (3.5.4), (3.5.5) we deduce the equality

$$\Psi(z) = -\frac{1}{\pi} \int\limits_0^\omega \frac{\ln t\, dt}{\sqrt{t(\omega-t)}(t-z)} - \int\limits_\omega^\infty \frac{dt}{\sqrt{t(t-\omega)}(t-z)}.$$

By the Sokhotski–Plemelj [55] formulas we have

$$\Psi_+(x) + \Psi_-(x) = -\frac{2}{\pi} \int_0^{\omega'} \frac{\ln t \, dt}{\sqrt{t(\omega - t)}(t - x)} - 2 \int_\omega^\infty \frac{dt}{\sqrt{t(t - \omega)}(t - x)}. \qquad (3.5.6)$$

According to (3.5.4), (3.5.5) we obtain

$$\Psi_+(x) + \Psi_-(x) = -\frac{2\pi}{\sqrt{x(\omega - x)}}, \qquad 0 < x < \omega. \qquad (3.5.7)$$

With the help of Euler substitution, we deduce the equality

$$\int \frac{dt}{(t - p)\sqrt{t(t - \omega)}} = \frac{1}{\sqrt{p(\omega - p)}} \arcsin \frac{(-\omega/2 + p)(t - p) - p(\omega - p)}{(t - p)\omega/2} + C,$$

where $|p| < \omega < t$. Thus, we have

$$\int_\omega^\infty \frac{dt}{\sqrt{t(t - \omega)}(t - x)} = \frac{1}{\sqrt{x(\omega - x)}} \left(\arcsin \frac{2x - \omega}{\omega} + \frac{\pi}{2} \right), \qquad 0 < x < \omega. \quad (3.5.8)$$

From (3.5.6)–(3.5.8) the equality (3.5.3) follows.   □

Theorem 3.5.2. *Let* $\mathcal{L}_1(x)$, $\mathcal{L}_2(x)$ *satisfy the relations*

$$S_n \mathcal{L}_k = x^{k-1} \quad (k = 1, 2), \quad \mathcal{L}_k \in L(0, \omega). \qquad (3.5.9)$$

If $n \geq 1$, *then we have*

$$\pi \int_x^\omega \mathcal{L}_k(t) P_m(t - x) \, dt = \sqrt{x(\omega - x)} M_{k,n-1}(x) + N_{k,m}(x) \left(\frac{\pi}{2} - \arcsin \frac{2x - \omega}{\omega} \right),$$

$$(3.5.10)$$

where $M_{k,n-1}(x)$ *and* $N_{k,m}(x)$ *are polynomials and*

$$\deg M_{k,n-1}(x) \leq n - 1, \qquad \deg N_{k,m}(x) \leq m.$$

Proof. Putting

$$X(z) = \sqrt{\frac{z - \omega}{z}} z, \qquad (3.5.11)$$

we write

$$X_\pm(t) = \pm i \sqrt{t(\omega - t)}, \quad 0 < t < \omega. \qquad (3.5.12)$$

As in the works [12, 17] we introduce the function

$$\Phi(z) = \int_0^\omega \left(P_m(t - z) \ln \left(1 - \frac{t}{z} \right) + Q_n(t - z) \right) f(t) \, dt, \qquad (3.5.13)$$

where $f(t) \in L(0, \omega)$. Limiting values $\Phi_{\pm}(x)$ on the segment $[0, \omega]$ are defined by the relations

$$\Phi_{\pm}(x) = \int_0^\omega \Big(P_m(t-x) \ln |t-x| + Q_n(t-x) \Big) f(t)\, dt$$

$$- \ln x \int_0^\omega P_m(t-x) f(t)\, dt \pm \pi i \int_x^\omega P_m(t-x) f(t)\, dt. \qquad (3.5.14)$$

Thus, the equalities

$$\Phi_+(x) + \Phi_-(x) = 2 \int_0^\omega \Big(P_m(t-x) \ln |t-x| + Q_n(t-x) \Big) f(x)\, dt$$

$$- 2 \ln x \int_0^\omega P_m(t-x) f(t)\, dt, \qquad (3.5.15)$$

$$\Phi_+(x) - \Phi_-(x) = 2\pi i \int_x^\omega P_m(t-x) f(t)\, dt \qquad (3.5.16)$$

are valid. Using the results of the theory of the singular integral equations (see [55, Chap. 5]), we obtain

$$\Phi(z) = \frac{1}{2\pi i} X(z) \int_0^\omega \frac{g(t)\, dt}{X_+(t)\, (t-z)} + X(z) R_{n-1}(z), \qquad (3.5.17)$$

where

$$g(x) = 2 \int_0^\omega \Big(P_m(t-x) \ln |t-x| + Q_n(t-x) \Big) f(t)\, dt - 2 \ln x \int_0^\omega P_m(t-x) f(t)\, dt,$$

$$\qquad (3.5.18)$$

$R_{n-1}(x)$ is a polynomial, $\deg R_{n-1}(x) \le n-1$. It follows from the Sokhotski–Plemelj formula [55, Chap. 2, §16] and from (3.5.17) that

$$\Phi_{\pm}(x) = \pm\sqrt{x(\omega - x)} \left(\pm \frac{g(x)}{2\sqrt{x(\omega - x)}} + \frac{1}{2\pi i} \int_0^\omega \frac{g(t)\, dt}{\sqrt{t(\omega - t)}(t-x)} + i R_{n-1}(x) \right).$$

$$\qquad (3.5.19)$$

Comparing (3.5.15) with (3.5.19) we have

$$\pi \int_x^\omega P_m(t-x) f(t)\, dt = \sqrt{x(\omega - x)} \left(R_{n-1}(x) - \frac{1}{2\pi} \int_0^\omega \frac{g(t)\, dt}{\sqrt{t(\omega - t)}(t-x)} \right).$$

$$\qquad (3.5.20)$$

From formulas (3.5.9), (3.5.18), (3.5.20) we deduce

$$\pi \int_x^\omega \mathcal{L}_k(t) P_m(t-x)\,dt = \sqrt{x(\omega-x)} \tag{3.5.21}$$

$$\times \left(R_{k,n-1}(x) - \int_0^\omega \frac{t^{k-1}\,dt}{\sqrt{t(\omega-t)}(t-x)} + \frac{1}{\pi}\int_0^\omega \frac{(\ln t) q_{k,m}(t)\,dt}{\sqrt{t(\omega-t)}(t-x)} \right), \quad k=1,2.$$

Here $R_{k,n-1}$ are certain polynomials, $\deg R_{k,n-1}(x) \le n-1$ and

$$q_{k,m}(t) = \int_0^\omega \mathcal{L}_k(s) P_m(s-t)\,ds.$$

We write now

$$\int_0^\omega \frac{\ln t\, q_{k,m}(t)\,dt}{\sqrt{t(\omega_t)}(t-x)} = \int_0^\omega \frac{\ln t\left(q_{k,m}(t) - q_{k,m}(x)\right)\,dt}{\sqrt{t(\omega-t)}(t-x)} + q_{k,m}(x)\int_0^\omega \frac{\ln t\,dt}{\sqrt{t(\omega-t)}(t-x)}. \tag{3.5.22}$$

It is easy to see that the first integral on the right-hand side of (3.5.22) is a polynomial with degree not larger than $(n-1)$. The assertion of the theorem follows directly from Lemma 3.5.1 and the equalities (3.5.21), (3.5.22). $\qquad\square$

Corollary 3.5.3. *If S_n $(n \ge 1)$ has the form*

$$S_n f = \int_0^\omega \left(\ln|x-t| + Q_n(t-x) \right) f(t)\,dt, \tag{3.5.23}$$

then the equality

$$\pi \int_x^\omega \mathcal{L}_k(t)\,dt = \sqrt{x(\omega-x)}\, M_{k,n-1}(x) + N_k \left(\frac{\pi}{2} - \arcsin\frac{2x-\omega}{\omega} \right), \tag{3.5.24}$$

where $M_{k,n-1}(x)$ are polynomials, $\deg M_{k,n-1}(x) \le n-1$ and $N_k = \displaystyle\int_0^\omega \mathcal{L}_k(t)\,dt$,

holds.

From (3.5.24) in the case $m=0$ we have

$$\mathcal{L}_k(x) = \frac{\varphi_k(x)}{\sqrt{x(\omega-x)}}, \quad k=1,2, \tag{3.5.25}$$

where $\varphi_k(x)$ are polynomials, $\deg \varphi_k(x) \leq n$. The relation (3.5.25) is valid not only for the operators of the form (3.5.23) but for other important cases as well.

The structure of the kernel $\Phi_1(x,t)$ of the operator T (see (2.2.31)) under the condition (3.5.25) is described in the following theorem.

Theorem 3.5.4. *If the relations* (3.5.25) *are valid and*

$$R = \int\limits_0^\omega \mathcal{L}_1(x)\, dx \neq 0$$

then

$$\Phi_1(x,t) = \sum_{k=0}^{n-2} B_k(x-t)\,(x+t-\omega)^{2k} 4\sqrt{xt(\omega-t)(\omega-x)}$$

$$+ B_{-1}(x-t)\ln \frac{\omega|x-t|}{\left(\sqrt{t(\omega-x)} + \sqrt{x(\omega-t)}\right)^2}, \qquad (3.5.26)$$

where $B_k(x)$ are polynomials, $\deg B_k(x) \leq 2(n-2-k)$.

Proof. By virtue of (2.2.26) we have

$$Q_1(x,t) = \frac{\varphi_1(\omega-t)\varphi_2(x) - \varphi_2(\omega-t)\varphi_1(x)}{R\sqrt{xt(\omega-t)(\omega-x)}}.$$

It follows from (2.2.30) that

$$\Phi_1(x,t) = \frac{2}{R} \int\limits_{x+t}^{2\omega-|x-t|} \frac{\varphi_2\left(\frac{s+x-t}{2}\right)\varphi_1\left(\omega - \frac{s-x+t}{2}\right) - \varphi_2\left(\omega - \frac{s-x+t}{2}\right)\varphi_1\left(\frac{s+x-t}{2}\right)}{\left(\left(s^2 - (x-t)^2\right)\left((2\omega-s)^2 - (x-t)^2\right)\right)^{1/2}}\, ds.$$

Introducing the variable $z = (\omega - s)^2$ instead of s, we obtain

$$\Phi_1(x,t) = \int\limits_{(|x-t|-\omega)^2}^{(\omega-(x+t))^2} \frac{\Psi(x,t,z)\, dz}{\sqrt{z^2 - 2z\left(\omega^2 + (x-t)^2\right) + \left(\omega^2 - (x-t)^2\right)^2}}, \qquad (3.5.27)$$

where

$$\Psi(x,t,z) = \frac{\varphi_2\left(\frac{\omega-\sqrt{z}+x-t}{2}\right)\varphi_1\left(\frac{\omega+\sqrt{z}+x-t}{2}\right) - \varphi_2\left(\frac{\omega+\sqrt{z}+x-t}{2}\right)\varphi_1\left(\frac{\omega-\sqrt{z}+x-t}{2}\right)}{\sqrt{z}R}.$$

We write $\varphi_1(x)$ and $\varphi_2(x)$ in the forms

$$\varphi_1\left(\frac{\omega+y}{2}\right) = \sum_{k=0}^n \alpha_k y^k, \quad \varphi_2\left(\frac{\omega+y}{2}\right) = \sum_{k=0}^n \beta_k y^k.$$

Then the function $\Psi(x, t, z)$ admits the representation

$$
\Psi(x, t, z) = \left(\sum_{k>p} \alpha_k \beta_k \left((x-t)^2 - z \right)^p \left((x - t + \sqrt{z})^{k-p} - (x - t - \sqrt{z})^{k-p} \right) \right.
$$

$$
\left. + \sum_{k<p} \alpha_k \beta_k \left((x-t)^2 - z \right)^k \left((x - t - \sqrt{z})^{p-k} - (x - t + \sqrt{z})^{p-k} \right) \right)
$$

$$
\times \frac{1}{\sqrt{z}R}. \qquad (3.5.28)
$$

Let us rewrite (3.5.28) in the form

$$
\Psi(x, t, z) = \sum_{k=0}^{n-1} z^k A_k(x - t), \qquad (3.5.29)
$$

where $A_k(x)$ are polynomials, $\deg A_k(x) \le 2(n - k - 1)$. Using Ostrogradski's method, from (3.5.27) and (3.5.29) we deduce (3.5.26).

The functions $B_k(x)$ are defined by the relations

$$
A_m(x) = m B_{m-1}(x) - (2m + 1) \left(\omega^2 + x^2 \right) B_m(x)
$$

$$
+ (m + 1) \left(\omega^2 - x^2 \right)^2 B_{m+1}(x) \quad (0 < m \le n - 1), \quad (3.5.30)
$$

$$
A_0(x) = - \left(\omega^2 + x^2 \right) B_0(x) + \left(\omega^2 - x^2 \right)^2 B_1(x) + B_{-1}(x). \qquad (3.5.31)
$$

Here we suppose that

$$
B_k(x) = 0, \quad \text{when} \ \ k > n - 2.
$$

From (3.5.30), (3.5.31) we deduce that $B_k(x)$ are polynomials of orders $\deg B_k(x) \le 2(n - 2 - k)$. $\qquad \square$

2. Let us consider the equation

$$
Sf = f(x) + \int_0^\omega k(x - t) f(t) \, dt = \varphi(x), \qquad (3.5.32)
$$

where the kernel $k(x)$ has the following structure

$$
k(x) = \sum_{s=1}^N \beta_s \, e^{-\alpha_s |x|}. \qquad (3.5.33)
$$

Equations of the form (3.5.32), (3.5.33) play an essential role in a number of problems (the problem of optimal prediction [112], the inverse spectral problem [90], nonlinear integrable equations [91]).

We suppose that the operator S is invertible. Then the operator S^{-1} has the form

$$S^{-1}\varphi = \varphi(x) + \int_0^\omega \gamma(x,t)\varphi(t)\,dt = f(x). \qquad (3.5.34)$$

It is known [101] (see also [5, 6, 56, 57]) that the kernel $\gamma(x,t)$ can be constructed explicitly.

3.6 On certain problems of hydrodynamics

1. Let us consider the equation

$$Sf = -\frac{1}{2\pi}\int_0^\omega \left(\ln|x-t| + r(x-t)\right) f(t)\,dt = \varphi(x), \qquad (3.6.1)$$

where the inequality

$$|r'(x)| \leq c \quad (-\omega \leq x \leq \omega) \qquad (3.6.2)$$

is valid for some c. Differentiating both sides of (3.6.1) we obtain a new equation

$$Kf = -\frac{1}{2\pi}\int_0^\omega \left(\frac{1}{x-t} + k(x-t)\right) f(t)\,dt = \varphi'(x), \qquad (3.6.3)$$

where

$$k(x) = r'(x). \qquad (3.6.4)$$

In their study of the motion of a wing under the surface of a fluid, Keldysh and Lavrent'ev [34] arrived at the equation (3.6.3) where

$$k(x,h,\nu) = \mathrm{Re}\left(\frac{1}{x-hi} - 2\nu i\exp(i\nu x)\int_{-\infty}^x \frac{e^{i\nu t}}{t-hi}\,dt\right), \quad h>0, \quad \nu>0. \quad (3.6.5)$$

Here $f(x)$ is the intensity of the turbulent layer and $\varphi(x)$ is the profile of the wing. Since (3.6.3) may have a set of solutions, we add the condition

$$f(x) = O(1), \quad x \to 0, \qquad (3.6.6)$$

which follows from the Chaplygin–Zhukovskii postulate on finite velocity of fluid at the trailing edge of the wing.

In this section we find conditions for the existence and uniqueness of the solution of (3.6.3) satisfying condition (3.6.6).

2. Let us study the behavior of the solution for the cases $x \to 0$ and $x \to \omega$.

Lemma 3.6.1. *Suppose that $k(x)$ and $\varphi'(x)$ belong to the Hölder class $H_{1/2}$. Then any solution of (3.6.3) that belongs to $L^p(0, \omega)$ should be continuous on $(0, \omega)$ and*

$$f(x) = \frac{\alpha_f}{\sqrt{x}} + O(1), \quad x \to 0; \qquad f(x) = \frac{\overset{\cdot}{\beta_f}}{\sqrt{\omega - x}} + O(1), \quad x \to \omega. \qquad (3.6.7)$$

Proof. Putting

$$g(x) = \varphi'(x) + \frac{1}{2\pi} \int_0^\omega k(x - t) f(t) \, dt, \qquad (3.6.8)$$

we rewrite (3.6.3) in the form

$$-\frac{1}{2\pi} \int_0^\omega \frac{1}{x - t} f(t) \, dt = g(x)$$

from which it follows that (see [55, Chap. 5])

$$f(x) = \frac{1}{\sqrt{x(\omega - x)}} \left(\frac{2}{\pi} \int_0^\omega \frac{\sqrt{t(\omega - t)}}{x - t} g(t) \, dt + c \right). \qquad (3.6.9)$$

By virtue of (3.6.8) and of the assumptions of the lemma, we have $g(x) \in H_{1/2}$. Hence, according to [55, Chap. 2], the integral on the right-hand side of (3.6.9) also belongs to $H_{1/2}$, and the statement of the lemma follows. $\square$

Theorem 3.6.2. *Suppose that the operator S has the form (3.6.1) and the equation*

$$Sf = 0 \qquad (3.6.10)$$

has in $L^p(0, \omega)$ $(1 < p < 2)$ only a trivial solution. If $k(x) = r'(x) \in H_{1/2}$ and

$$r(x) = \overline{r(-x)} \quad (0 \le x \le \omega), \qquad (3.6.11)$$

then (3.6.3) has (in the space in $L^p(0, \omega)$) one and only one solution satisfying (3.6.6).

Proof. By Corollary 3.3.2, equation (3.6.1) has the unique solution $f(x) = \mathcal{L}_\varphi$. Then the general solution of (3.6.3) has the form

$$f(x) = \mathcal{L}_\varphi + c\mathcal{L}_1, \qquad (3.6.12)$$

where c is an arbitrary constant. Lemma 3.6.1 implies that

$$\mathcal{L}_1(x) = \frac{\alpha_1}{\sqrt{x}} + O(1), \quad x \to 0; \qquad \mathcal{L}_1(x) = \frac{\beta_1}{\sqrt{\omega - x}} + O(1), \quad x \to \omega, \quad (3.6.13)$$

$$\mathcal{L}_\varphi(x) = \frac{\alpha_\varphi}{\sqrt{x}} + O(1), \quad x \to 0; \qquad \mathcal{L}_\varphi(x) = \frac{\beta_\varphi}{\sqrt{\omega - x}} + O(1), \quad x \to \omega. \quad (3.6.14)$$

From part I of Theorem 2.2.4 and from (3.6.13) it follows that

$$|\alpha_1| + |\beta_1| \neq 0. \tag{3.6.15}$$

Since (3.6.11) yields $\mathcal{L}_1(x) = \overline{\mathcal{L}_1(\omega - x)}$, we have $\alpha_1 = \overline{\beta}_1$. Then (3.6.15) implies that

$$\alpha_1 \neq 0. \tag{3.6.16}$$

Putting $c = -\dfrac{\alpha_\varphi}{\alpha_1}$ in (3.6.12) we recover the unique solution of (3.6.3) that satisfies (3.6.6). $\qquad\qquad\square$

Remark 3.6.3. The function $B(x, \lambda)$ has the following properties

$$B(x, \lambda) = \begin{cases} \dfrac{X(\lambda)}{\sqrt{x}} + O(1), & x \to 0, \\[3mm] \dfrac{Y(\lambda)}{\sqrt{\omega - x}} + O(1), & x \to \omega. \end{cases} \tag{3.6.17}$$

The introduced functions $X(\lambda)$ and $Y(\lambda)$ describe the behavior of $B(x, \lambda)$ on the ends of interval. So, these functions are similar to the Ambartsumian functions (1.3.13).

The results of Subsection 4 of Section 2.2, have the following interpretation in the case of the equation of motion of a wing: from the intensities $\alpha\mathcal{L}_1(x) - \beta\mathcal{L}_2(x)$ of the turbulent layer for the simplest profiles of parabolic form $\varphi'(x) = \alpha - \beta x$ one can recover explicitly the intensity $f(x)$ of the turbulent layer for any sufficiently smooth profile $\varphi(x)$.

3. It was assumed in Section 3.2 that equation (3.6.10) has only the trivial solution in $L^p(0, \omega)$. Now we formulate conditions for this to be valid. We introduce the family of operators

$$S_h f = -\frac{1}{2\pi} \int_0^\omega f(t)\big(\ln|x - t| + r(x - t, h)\big)\, dt, \quad 0 < h < \infty, \tag{3.6.18}$$

where

$$\left| \frac{\partial}{\partial x} r(x, h) \right| + \left| \frac{\partial}{\partial h} r(x, h) \right| \leq c(h), \quad r(x, h) = \overline{r(-x, h)}. \tag{3.6.19}$$

It follows from (3.6.19) that $S_h f \in L^q(0, \omega)$. Hence, the scalar product

$$\langle S_h f, f \rangle = \int_0^\omega \varphi(x)\overline{f(x)}\, dx, \quad \text{where } S_h f = \varphi(x)$$

is well defined. Now, $\langle S_h f, f \rangle$ is a real number because $r(x, h) = \overline{r(-x, h)}$. The subspace H_m is said to be negative with respect to S_h if $\langle S_h f, f \rangle < 0$ for $f \in H_m$, $\|f\|_p \neq 0$. By $m(h)$ we denote the dimension of a maximal negative subspace for S_h. By virtue of (3.6.19), the operator $\dfrac{\partial S_h}{\partial h}$, which maps $L^p(0, \omega)$ into $L^q(0, \omega)$ is well defined. We assume further that

$$\frac{\partial S_h}{\partial h} < 0, \quad 0 < h < \infty. \tag{3.6.20}$$

Then $m(h)$ increases monotonically. Next, we put

$$m_0 = \lim_{h \to +0} m(h), \quad m_\infty = \lim_{h \to +\infty} m(h).$$

If

$$m_\infty < \infty, \tag{3.6.21}$$

then the monotonically increasing integer-valued function $m(h)$ has finitely many points h_k of discontinuity: $0 < h_1 < h_2 < \cdots < h_n$ ($n \leq m_\infty - m_0$).

Theorem 3.6.4. *Suppose that* (3.6.18)–(3.6.21) *holds and*

$$h \neq h_k \quad (1 \leq k \leq n), \qquad \varphi'(x) \in L^{p_1}(0, \omega) \quad (1 < p < p_1 \leq 2).$$

Then

$$S_h f = \varphi \tag{3.6.22}$$

has one and only one solution in $L^p(0, \omega)$.

Proof. We write $\mathcal{M}_-(h)$ for the space spanned by the eigenfunctions of S_h corresponding to negative eigenvalues. The following statements are easily checked:

   I. $\mathcal{M}_-(h)$ belongs to $L^q(0, \omega)$.

  II. $\mathcal{M}_-(h)$ is negative with respect to S_h.

 III. $\dim \mathcal{M}_h = m(h)$.

Suppose that for some h_0

$$S_{h_0} f = 0 \tag{3.6.23}$$

has a non-trivial solution f_0 in $L^p(0, \omega)$. Then the subspace $\mathcal{M}(h_0)$, spanned by f_0 and $\mathcal{M}_-(h_0)$, is nonpositive with respect to S_{h_0}, and $\dim \mathcal{M}(h_0) = m(h_0) + 1$. Hence, by virtue of (3.6.20), $\mathcal{M}(h_0)$ is negative with respect to S_h for $h > h_0$, that is, from I–III we deduce that

$$m(h) \geq m(h_0) + 1, \quad h > h_0. \tag{3.6.24}$$

It follows from (3.6.24) that $m(h)$ is discontinuous at h_0. Thus, h_0 is one of the points h_k ($1 \leq k \leq n$). For $h \neq h_k$ the equation (3.6.23) has only a trivial solution. Now, Theorem 3.6.4 follows from Corollary 3.3.2. $\qquad\square$

4. Next, we study some specific integral equations which appear in problems of hydrodynamics.

Example 3.6.5. Consider the equation

$$S_h f = -\frac{1}{2\pi} \int_0^\omega f(t) \left(\ln \left(\frac{b}{2} |x - t| \right) + \frac{1}{2} \ln \left((x - t)^2 + h^2 \right) \right) dt = \varphi(x), \quad (3.6.25)$$

where $0 < h < \infty$, $b > 0$. It is easy to see that

$$\frac{\partial S_h}{\partial h} f = -\frac{h}{2\pi} \int_0^\omega f(t) \frac{1}{(x - t)^2 + h^2} \, dt. \tag{3.6.26}$$

Since the equality

$$\frac{2h}{x^2 + h^2} = \int_{-\infty}^{\infty} e^{ixt} e^{-|t|/h} \, dt \tag{3.6.27}$$

holds, the operator $\dfrac{\partial S_h}{\partial h}$ is negative. From (3.6.25) we deduce the asymptotic relation

$$\langle S_h f, f \rangle = \langle \widetilde{S} f, f \rangle - \frac{1}{2\pi} (\ln h) \, |\langle f, \mathbb{1} \rangle|^2 + o(1), \quad h \to \infty, \tag{3.6.28}$$

where

$$\widetilde{S} f = -\frac{1}{2\pi} \int_0^\omega f(t) \ln \left(\frac{b}{2} |x - t| \right) dt. \tag{3.6.29}$$

The maximal dimension of a negative subspace of $\widetilde{S}$ is denoted by m. It is easy to check the following statements.

   I. If $\omega \leq 2b$, then $m = 0$.
  II. If $\omega > 2b$, then $m = 1$.
 III. If $\langle f, \mathbb{1} \rangle = 0$, then $\langle \widetilde{S} f, f \rangle \geq 0$.

It follows from II–III and (3.6.28) that in this example we have

$$m_0 = m_\infty = 1 \text{ for } \omega \geq 2\sqrt{2b}. \tag{3.6.30}$$

From Theorem 3.6.4 and (3.6.30) we obtain the following corollary.

Corollary 3.6.6. *Suppose that* $\omega \geq 2\sqrt{2b}$ *and* $\varphi'(x) \in L^{p_1}(0, \omega)$ $(1 < p < p_1 \leq 2)$. *Then equation (3.6.25) has a unique solution in* $L^p(0, \omega)$.

Example 3.6.7. Consider the equation

$$S_h f = -\frac{1}{2\pi} \int\limits_0^\omega f(t) \left(\ln |x - t| - \frac{1}{2} \ln\big((x - t)^2 + h^2\big) \right) dt = \varphi(x). \qquad (3.6.31)$$

Now it follows from (3.6.26) that

$$\frac{\partial S_h}{\partial h} > 0. \qquad (3.6.32)$$

Since S_0 maps the whole space into zero, we deduce from (3.6.32) that

$$S_h > 0, \quad 0 < h < \infty. \qquad (3.6.33)$$

Hence, the equation $S_h f = 0$ has only a trivial solution in $L^p(0, \omega)$. From Corollary 3.3.2 we obtain another corollary.

Corollary 3.6.8. *If $\varphi'(x) \in L^{p_1}(0, \omega)$ $(1 < p < p_1 \leq 2)$, then equation (3.6.31) has a unique solution in $L^p(0, \omega)$.*

Example 3.6.9. The equation below is closely connected with Examples 3.6.5 and 3.6.7:

$$K_h f = -\frac{1}{2\pi} \int\limits_0^\omega f(t) \left(\frac{1}{x - t} + \frac{2b(x - t)}{(x - t)^2 + h^2} \right) dt = \varphi'(x), \quad b = \pm\frac{1}{2}. \qquad (3.6.34)$$

This equation represents the limiting cases of the Keldysh–Lavrent'ev equation when $\nu \to 0$ $(b = 1/2)$ and $\nu \to \infty$ $(b = -1/2)$. For $b = 1/2$ we choose a number ω such that $\omega \geq 2(2b)^{1/2}$. From Theorem 3.6.2 and Corollaries 3.6.6 and 3.6.8 we now deduce the following corollary.

Corollary 3.6.10. *If $\varphi'(x) \in H_{1/2}$, then for any h $(0 < h < \infty)$ the equation (3.6.34) has one and only one solution, which satisfies the Chaplygin–Zhukovskii condition (3.6.6), in $L^p(0, \omega)$.*

Example 3.6.11 (Keldysh–Lavrent'ev equation). It is easy to see that Corollary 3.6.10 remains valid for the Keldysh–Lavrent'ev equation (3.6.3), (3.6.5) for sufficiently small and for sufficiently large ν $(\nu > 0)$. Using properties of analytic operator-valued functions we can refine this result.

Corollary 3.6.12. *If $\varphi'(x) \in H_{1/2}$, then for any h $(0 < h < \infty)$ and any ν $(\nu > 0)$ with finitely many exceptions $(\nu \neq \nu_k, 1 \leq k \leq N(h))$, the Keldysh–Lavrent'ev equation has one and only one solution satisfying (3.6.6).*

Example 3.6.13. The equation

$$S\mathcal{L}_1 = \int\limits_0^\omega \mathcal{L}_1(t) \int\limits_1^\infty \cos\left(\nu\lambda(x - t)\right) \frac{d\lambda}{\sqrt{\lambda^2 - 1}} \, dt = 1, \quad \nu > 0, \qquad (3.6.35)$$

is used in problems on the optimal configuration of a ship (see [37]). Since S is positive and

$$\int\limits_{1}^{\infty} \cos(\nu \lambda x) \frac{\mathrm{d}\lambda}{\sqrt{\lambda^2 - 1}} = -\ln|x| + k(x), \quad k'(x) \in H_{1/2},$$

the equation (3.6.35) has in $L^p(0, \omega)$ a unique continuous in $(0, \omega)$ solution $\mathcal{L}_1(x, \nu)$ and this solution satisfies the asymptotic relations

$$\mathcal{L}_1(x, \nu) = \begin{cases} \dfrac{\alpha_1(\nu)}{\sqrt{x}} + O(1) & (x \to 0), \\ \dfrac{\alpha_1(\nu)}{\sqrt{\omega - x}} + O(1) & (x \to \infty). \end{cases} \tag{3.6.36}$$

From the analyticity of $\alpha_1(\nu)$, M.G. Kreĭn deduced the inequality

$$\alpha_1(\nu) \neq 0 \tag{3.6.37}$$

for all $\nu > 0$ with finitely many exceptions. He formulated the problem whether (3.6.37) holds for all $\nu > 0$.

Recall that solutions of the form (3.6.13) (and $\mathcal{L}_1(x, \nu)$, in particular) satisfy (3.6.15). Formula (3.6.15) resolves Kreĭn's problem. We obtain the following result.

Corollary 3.6.14. *The inequality* (3.6.37) *holds for all* $\nu > 0$.

This result means that the extremal problem above has no solution in the class of possible configurations of a ship.

3.7   Equations from the contact theory of elasticity

1. Suppose that two contacting elastic bodies are bounded before pressure in the xy-plane by the curves

$$y = u(x), \quad y = v(x).$$

Suppose further that the contact of the compressed elastic bodies takes place along the interval $[0, \omega]$. Then the contact strain $f(x)$ is a solution of the integral equation [110], [20, Chap. 6]

$$Sf = -\int\limits_{0}^{\omega} f(t) \Big(\ln|x - t| + r(x - t) \Big) \, \mathrm{d}t = \varphi(x), \tag{3.7.1}$$

where, for some c,

$$|r'(x)| \leq c, \quad -\omega \leq x \leq \omega. \tag{3.7.2}$$

The right-hand side $\varphi(x)$ can be expressed in terms of $u(x)$ and $v(x)$ by the formula

$$\varphi(x) = \frac{1}{\theta}\left(\theta_1 - u(x) - v(x)\right), \tag{3.7.3}$$

where $\theta \neq 0$ and θ_1 are constants. If $f \in L^p(0,\omega)$ $(1 < p < 2)$ then, by virtue of (3.7.1) and (3.7.2), $Sf \in L^q(0,\omega)$. Hence, the scalar product $\langle Sf, g \rangle$ for $f, g \in L^p(0,\omega)$ is well defined. The operators occurring in various problems of the contact theory of elasticity are positive. From Corollary 3.3.2 we have immediately the following result.

Theorem 3.7.1. *Let S be a positive operator of the form* (3.7.1), (3.7.2). *Then for $\varphi'(x) \in L^{p_1}(0,\omega)$ $(1 < p < p_1 \leq 2)$ the equation* (3.7.1) *has one and only one solution in $L^p(0,\omega)$.*

We denote by $f_1(x)$ and $f_2(x)$ the contact strain for the simplest plates $u_1(x) = 1$ and $u_2(x) = x$ pressed into the elastic base $y = 0$. From Theorem 3.7.1 it follows (for $\theta_1 \neq 1$) that

$$f_1(x), \; f_2(x) \in L^p(0,\omega), \quad \int_0^\omega f_1(t)\,dt \neq 0. \tag{3.7.4}$$

It is easy to see that

$$f_1(x) = \frac{1}{\theta}(\theta_1 - 1)\mathcal{L}_1, \quad f_2(x) = \frac{1}{\theta}(\theta_1\mathcal{L}_1 - \mathcal{L}_2). \tag{3.7.5}$$

Then the following statement is valid:
If $u'(x)$, $v'(x) \in L^{p_1}(0,\omega)$ $(1 < p < p_1 \leq 2)$, then the contact strain $f(x)$ is explicitly determined by the contact strains $f_1(x)$ and $f_2(x)$.

Indeed, using (3.7.5), we recover the functions $\mathcal{L}_1$ and $\mathcal{L}_2$ from f_1 and f_2. It was shown in Section 2.2 that f can be expressed explicitly in terms of $\mathcal{L}_1$ and $\mathcal{L}_2$, which proves the statement.

The interpretation of the results of Section 2.2 from the point of view of contact problems was suggested to the author by M.G. Kreĭn.

2. Let us consider the equation

$$Sf = \frac{1}{\pi}\int_0^\omega \left(\frac{1}{x-t} + \alpha e^{-\xi(x-t)} \int_{-\infty}^{x-t} (e^{\xi\tau}/\tau)d\tau \right) f(t)\,dt = \varphi(x). \tag{3.7.6}$$

Equation (3.7.6) plays an essential role in a number of problems in the theory of elasticity (see [11]). Using direct calculations, A.G. Buslaev [11] proved that the functions $\mathcal{L}_1(x)$ and $\mathcal{L}_2(x)$ for the equation (3.7.6) have, respectively, the forms

$$\mathcal{L}_1(x) = -\frac{\xi}{\xi+\alpha}\frac{1}{\sqrt{x(\omega-x)}}\left(\frac{\omega}{2}c(z)+x-\frac{\omega}{2}\right), \qquad z = \frac{\xi\omega}{2}, \tag{3.7.7}$$

$$\mathcal{L}_2(x) = -\frac{\xi}{\xi+\alpha}\frac{1}{\sqrt{x(\omega-x)}}\left(\frac{\omega^2}{4}(c(z)-1) - \frac{1}{\xi+\alpha}\frac{\omega}{2}c(z)\right.$$

$$\left. + \left(\frac{\omega}{2} + \frac{\alpha}{(\xi+\alpha)\xi}\right)\left(x - \frac{\omega}{2}\right) + \left(x - \frac{\omega}{2}\right)^2\right), \tag{3.7.8}$$

where $c(z) = K_1(z)/K_0(z)$ and $K_\nu(z)$ are modified Bessel functions ($\xi + \alpha > 0$, $0 < \omega < \infty$). Hence, the numbers

$$R = \int_0^\omega \mathcal{L}_1(t)\,dt = -\frac{\xi}{\xi+\alpha}\frac{\pi\omega}{2}c(z) \tag{3.7.9}$$

and

$$\gamma = \frac{1}{R}\int_0^\omega \left(t\mathcal{L}_1(\omega - t) - \mathcal{L}_2(t)\right)dt = \frac{1}{\xi+\alpha} \tag{3.7.10}$$

are not equal to zero. Thus, the operator T_γ can be constructed via the formulas (2.2.17) and (2.2.29).

We note that the functions $\mathcal{L}_k(x)$ ($k = 1, 2$) defined by the formulas (3.7.7), (3.7.8) have the forms

$$\mathcal{L}_k(x) = \varphi_k(x)\big/\sqrt{x(\omega - x)}, \tag{3.7.11}$$

where $\varphi_k(x)$ are polynomials and $\deg\varphi_k \le 2$. From the results of Section 3.5 we have

$$\Phi_1(x,t) = 4\,B_0(x-t)\sqrt{xt(\omega-x)(\omega-t)}$$

$$+ B_{-1}(x-t)\ln\frac{\omega|x-t|}{\left(\sqrt{x(\omega-t)} + \sqrt{t(\omega-x)}\right)^2}, \tag{3.7.12}$$

where the coefficients $B_k(x)$ are defined with the help of the system of equations

$$\begin{cases} B_0(x) = A_1(x), \\ B_-(x) - (\omega^2 + x^2)B_0(x) = A_0(x). \end{cases} \tag{3.7.13}$$

Here, according to Section 3.5, the equalities

$$A_0(x) = -\frac{\xi^2}{R(\xi+\alpha)^2}\left(\frac{\omega^2}{4} + \frac{\omega}{2\gamma}c(z)x + \frac{1}{4}x^2\right), \tag{3.7.14}$$

$$A_1(x) = \frac{\xi^2}{4R(\xi+\alpha)^2} \tag{3.7.15}$$

are valid. Therefore, from(3.7.13) we obtain

$$B_0(x) = \frac{\xi^2}{4R(\xi + \alpha)^2}, \quad B_-(x) = -\frac{\xi^2}{R(\xi + \alpha)^2} \frac{\omega}{2} c(z) \left(\frac{1}{\xi} + x\right). \quad (3.7.16)$$

Using formulas (2.2.17), (2.2.29) and (3.7.12), (3.7.16) the operators C_γ, T and T_γ can be constructed explicitly.

3.8 The equation of radiation transfer

The equation of radiation transfer in a planar medium has the form [29, 100]

$$\eta \frac{\partial J(\tau, \eta)}{\partial \tau} = -J(\tau, \eta) + \varepsilon(\tau), \quad (3.8.1)$$

where $J(\tau, \eta)$ is the total intensity of radiation,

$$\varepsilon(\tau) = g(\tau) + \frac{1}{2}\lambda \int_{-1}^{1} J(\tau, \eta')\, d\eta' \quad (3.8.2)$$

is a sourse function and $g(\tau)$ is the intensity of intrinsic isotropic radiation of the medium, $\arccos \eta$ is the angle between the selected direction of radiation and the external normal to the $\tau = 0$.

The boundary conditions are written in the following form (see [100, 109]):

$$J(0, \eta) = \varphi_1(\eta), \quad \eta < 0; \qquad J(\tau_0, \eta) = \varphi_2(\eta), \quad \eta > 0. \quad (3.8.3)$$

The equation (3.8.1) with boundary conditions (3.8.3) can be written in the integral form

$$J(\tau, \eta) = \begin{cases} \int_\tau^{\tau_0} \varepsilon(\tau') \exp\left(\frac{\tau - \tau'}{\eta}\right) \frac{d\tau'}{\eta} + \varphi_2(\eta) \exp\left(\frac{\tau - \tau_0}{\eta}\right), & \eta > 0, \\[2mm] -\int_0^\tau \varepsilon(\tau') \exp\left(\frac{\tau - \tau'}{\eta}\right) \frac{d\tau'}{\eta} + \varphi_1(\eta) \exp\left(\frac{\tau}{\eta}\right), & \eta < 0. \end{cases} \quad (3.8.4)$$

Substituting expression (3.8.4) into (3.8.2), we obtain the equation

$$\varepsilon(\tau) = \int_0^{\tau_0} k\left(|\tau - \tau'|\right) \varepsilon(\tau')\, d\tau' + g(\tau), \quad (3.8.5)$$

where

$$k(x) = \frac{1}{2}\lambda E_1(x), \quad E_k(x) = \int_0^1 e^{-x/\eta} \eta^{k-2}\, d\eta. \quad (3.8.6)$$

The functions N_1 and N_2 are defined by the equalities

$$SN_1 = \frac{1}{2}\lambda E_2(\tau), \quad SN_2 = \mathbb{1}, \tag{3.8.7}$$

where the operator S has the form

$$Sf = f(\tau) - \int_0^{\tau_0} k\big(|\tau - \tau'|\big) f(\tau') \,d\tau'. \tag{3.8.8}$$

Thus, the source function $\varepsilon(\tau)$ in the general case can be constructed explicitly, if the source functions $\varepsilon_I(\tau)$ and $\varepsilon_{II}(\tau)$ for the simplest cases

  I. $g(\tau) = \dfrac{1}{2}\lambda E_2(\tau)$,

  II. $g(\tau) = \mathbb{1}$

are known.

Chapter 4

Eigensubspaces and Fourier Transform

Consider a bounded operator S (in the space L^p) of the form

$$Sf = \frac{\mathrm{d}}{\mathrm{d}x} \int_0^\omega s(x-t)f(t)\,\mathrm{d}t, \quad s(x) \in L^q(-\omega,\omega), \tag{4.0.1}$$

where $\frac{1}{p} + \frac{1}{q} = 1$. By H_ν we denote the subspace of eigenvectors f of S corresponding to the eigenvalue ν, that is, the subspace of f such that

$$Sf = \nu f, \quad f \in L^p(0,\omega). \tag{4.0.2}$$

In this chapter, we study the structure of H_ν under the assumption that

$$1 \le \dim H_\nu < \infty. \tag{4.0.3}$$

For the case $p = 2$, $S = S^*$ we investigate the distribution of the roots of the function

$$\mathcal{D}(z) = \int_0^\omega e^{izt}\overline{\varphi(t)}\,\mathrm{d}t, \quad \varphi(x) \in H_\nu. \tag{4.0.4}$$

In this way, we generalize the results by M.G. Kreĭn (see [42]) for the operator

$$Sf = \alpha f + \int_0^\omega k(x-t)f(t)\,\mathrm{d}t \tag{4.0.5}$$

where $\alpha = \overline{\alpha}$, $k(x) = \overline{k(-x)} \in L(0,\omega)$. In order to illustrate our general theorems we consider the operator

$$Sf = \int_{-1}^1 \frac{\sin c(x-t)}{\pi(x-t)} f(t)\,\mathrm{d}t, \quad c > 0, \tag{4.0.6}$$

which plays an essential role in the problems of radioelectronics [15, Chap. 6].

4.1   Classification of eigensubspaces

Definition 4.1.1. Let $\dim H_\nu = n$ $(1 \leq n < \infty)$. We say that the subspace H_ν belongs to the class $\mathcal{H}_+$, if there is such a basis $\{\varphi_k\}$ $(1 \leq k \leq n)$ and such a set of n numbers α_k, that the equality

$$\int_0^\omega \varphi_k(x)\big(N(x) + \alpha_k\big)\,\mathrm{d}x = 0, \quad N(x) = -s(-x) \tag{4.1.1}$$

holds. We say that $H_\nu \in \mathcal{H}_-$ if, for all $\varphi \in H_\nu$, we have $\langle \varphi, \mathbb{1}\rangle = 0$, that is,

$$\int_0^\omega \varphi(x)\,\mathrm{d}x = 0. \tag{4.1.2}$$

Remark 4.1.2. If the conditions

$$\langle \varphi_k, \mathbb{1}\rangle \neq 0 \quad (1 \leq k \leq n) \tag{4.1.3}$$

are valid, relations (4.1.1) hold, and so $H_\nu \in \mathcal{H}_+$.

Lemma 4.1.3. *Assume that* $1 \leq \dim H_\nu < \infty$. *Then the subspace* H_ν *cannot belong to* $\mathcal{H}_+$ *and* $\mathcal{H}_-$ *simultaneously.*

Proof. Let H_ν belong to $\mathcal{H}_+$ and to $\mathcal{H}_-$. Then formulas (4.1.1) and (4.1.2) imply that $\langle \varphi, \mathbb{1}\rangle = 0$ and $\langle \varphi, \overline{N}\rangle = 0$. Since $\langle \varphi, \mathbb{1}\rangle = 0$, we obtain that $(A - A^*)\varphi = 0$, that is,

$$AS\varphi = \nu A\varphi = \nu A^*\varphi. \tag{4.1.4}$$

In view of $\langle \varphi, \mathbb{1}\rangle = 0$ and $\langle \varphi, \overline{N}\rangle = 0$, formula (1.1.10) yields

$$(AS - SA^*)\varphi = 0. \tag{4.1.5}$$

From (4.1.4) and (4.1.5) we derive

$$(S - \nu I)A^*\varphi = 0, \quad \varphi \in H_\nu. \tag{4.1.6}$$

Thus, H_ν is an invariant subspace of A^*. However, the finite-dimensional invariant subspaces of A^* do not exist, and we arrive at a contradiction, which proves the lemma. $\qquad\square$

At the end of this section, in Proposition 4.1.6, we show that each H_ν belongs to either $\mathcal{H}_+$ or $\mathcal{H}_-$.

According to Proposition 2.1.1, the operator S^* can be extended onto $L^p(0,\omega)$, and using (2.1.3) we see that

$$(S^* - \overline{\nu}I) = U(S - \nu I)U. \tag{4.1.7}$$

Moreover, taking adjoint operators to both parts of (1.1.10) and recalling that $A - A^* = i \int_0^\omega \cdot \, dt$, we obtain

$$\Big((S^* - \bar\nu I) A^* - A(S^* - \bar\nu I) \Big) f = -i \int_0^\omega \Big(\overline{M(t)} + \overline{N(x)} - \bar\nu \Big) f(t)\, dt. \qquad (4.1.8)$$

Setting $\varphi^* = U\varphi$, from (4.1.7) and (4.1.8) we deduce

$$i\,(S^* - \bar\nu I) A^* \varphi^* = \int_0^\omega \Big(\overline{M(t)} + \overline{N(x)} - \bar\nu \Big) \varphi^*(t)\, dt. \qquad (4.1.9)$$

Now, let us consider the case $H_\nu \in \mathcal{H}_+$. Lemma 4.1.3 implies that there is a function $\varphi(x) \in H_\nu$, such that $\langle \varphi, \mathbb{1} \rangle \neq 0$, and so $\langle \varphi^*, \mathbb{1} \rangle = \overline{\langle \varphi, \mathbb{1} \rangle} \neq 0$. Therefore, we rewrite (4.1.9) in the form

$$\frac{i\,(S^* - \bar\nu I) A^* \varphi^*}{\langle \varphi^*, \mathbb{1} \rangle} = \overline{N(x)} + \bar\alpha_\nu, \qquad \alpha_\nu = \frac{\langle M - \nu \mathbb{1}, \varphi^* \rangle}{\langle \varphi^*, \mathbb{1} \rangle}. \qquad (4.1.10)$$

Introduce the functions

$$\Phi(x) = \int_x^\omega \frac{\varphi(t)\, dt}{\langle \varphi, \mathbb{1} \rangle}, \qquad \Psi(x) = \int_x^\omega \frac{\varphi^*(t)\, dt}{\langle \varphi^*, \mathbb{1} \rangle}. \qquad (4.1.11)$$

From (4.1.10) it follows that

$$(S^* - \bar\nu I)\Psi = \overline{N(x)} + \bar\alpha_\nu. \qquad (4.1.12)$$

Besides (4.1.8), formula (1.1.10) also yields the identity

$$\Big((S - \nu I) A^* - A(S - \nu I) \Big) f = -i \int_0^\omega \Big(M(x) + N(t) - \nu \Big) f(t)\, dt. \qquad (4.1.13)$$

Hence, in a way which is similar to the proof of (4.1.12), we can show that

$$(S - \nu I)\Phi = M(x) + \beta_\nu, \qquad \beta_\nu = \frac{\langle N - \nu \mathbb{1}, \overline{\varphi} \rangle}{\langle \varphi, \mathbb{1} \rangle}. \qquad (4.1.14)$$

The operators $\dot B$ and C, acting from $L^p(0,\omega)$ into $L^q(0,\omega)$, are introduced by the formulas

$$B = A + P, \quad C = A + Q, \qquad (4.1.15)$$

$$Pf = -i\langle f, \Psi \rangle, \quad Qf = -i\langle f, \Phi \rangle, \quad f \in L^p(0,\omega). \qquad (4.1.16)$$

Since $\Psi, \Phi \in L^q(0, \omega)$, the operators above are well defined. The adjoint operators B^* and C^* also act from $L^p(0, \omega)$ into $L^q(0, \omega)$. From (1.1.10), (4.1.11), (4.1.12) and (4.1.14) it follows that

$$\left(B(S - \nu I) - (S - \nu I)C^* \right) f = -\mathrm{i}(\alpha_\nu + \beta_\nu + \nu) \int_0^\omega f(t)\, \mathrm{d}t. \qquad (4.1.17)$$

In view of $\varphi \in H_\nu$, relations (4.1.7) and $\varphi^* = U\varphi$ imply that

$$\left\langle \left(B(S - \nu I) - (S - \nu I)C^* \right)\varphi, \varphi^* \right\rangle = 0. \qquad (4.1.18)$$

Comparing (4.1.17) and (4.1.18) we deduce that

$$\alpha_\nu + \beta_\nu + \nu = 0.$$

Thus, the identity (4.1.17) takes the form

$$B(S - \nu I) - (S - \nu I)C^* = 0. \qquad (4.1.19)$$

Corollary 4.1.4. *Let $H_\nu \in \mathcal{H}_+$ ($\dim H_\nu < \infty$). Then H_ν is an invariant subspace of the operator C^*, where C is given by the second equalities in (4.1.15) and (4.1.16).*

Now, consider $H_\nu \in \mathcal{H}_-$. Using again Lemma 4.1.3, we derive that there is a function $\varphi(x) \in H_\nu$ such that

$$\langle \varphi, \overline{N} \rangle \neq 0. \qquad (4.1.20)$$

Putting

$$\Phi(x) = \int_x^\omega \frac{\varphi(t)\, \mathrm{d}t}{\langle \varphi, \overline{N} \rangle}, \quad \Psi(x) = \overline{\Phi(\omega - x)}, \qquad (4.1.21)$$

and taking into account (4.1.2), (4.1.13) and (2.1.3), we deduce that

$$(S - \nu I)\Phi = \mathbb{1}, \quad (S^* - \overline{\nu} I)\Psi = \mathbb{1}. \qquad (4.1.22)$$

Introduce B and C by (4.1.15), where P and Q are given by

$$Pf = -\mathrm{i}\langle f, \Psi \rangle \left(M(x) - \frac{\nu}{2} \right), \quad Qf = -\mathrm{i}\langle f, \Phi \rangle \left(\overline{N(x)} - \frac{\overline{\nu}}{2} \right). \qquad (4.1.23)$$

For $H_\nu \subset \mathcal{H}_-$ and operators B and C given by (4.1.15) and (4.1.23), formulas (4.1.13), (4.1.21) and (4.1.22) imply that (4.1.19) is valid again. The following corollary is immediate from (4.1.19).

Corollary 4.1.5. *Let $H_\nu \in \mathcal{H}_-$ ($\dim H_\nu < \infty$). Then H_ν is an invariant subspace of the operator C^*, where C is given by the second equalities in (4.1.15) and (4.1.23).*

Proposition 4.1.6. *Assume that* $1 \le \dim H_\nu < \infty$. *Then the subspace* H_ν *always belongs to either* $\mathcal{H}_+$ *or* $\mathcal{H}_-$ *but cannot belong to* $\mathcal{H}_+$ *and* $\mathcal{H}_-$ *simultaneously.*

Proof. According to Lemma 4.1.3, H_ν cannot belong to $\mathcal{H}_+$ and $\mathcal{H}_-$ simultaneously and it remains to prove the first part of the proposition's statement, namely, that H_ν always belongs to either $\mathcal{H}_+$ or $\mathcal{H}_-$.

Recall that if $r < \infty$ then there is a basis of H_ν of the form (see (2.2.35))

$$\varphi_k = (A^*)^{k-1}\varphi_1 \quad (1 \le k \le r). \tag{4.1.24}$$

By virtue of (4.1.13) the following relations are valid:

$$\int_0^\omega \varphi_k(x)\,\mathrm{d}x = 0 \quad \text{and} \quad \int_0^\omega \varphi_k(x)N(x)\,\mathrm{d}x = 0 \quad \text{for} \;\; 1 \le k < r. \tag{4.1.25}$$

Indeed, in view of (4.1.24) we have φ_k, $A^*\varphi_k \in H_\nu$ for $k < r$, and so (4.1.13) yields

$$\int_0^\omega \Big(M(x) + N(t) - \nu\Big)\varphi_k(t)\,\mathrm{d}t = 0. \tag{4.1.26}$$

Formula (4.1.26) can be rewritten in the form

$$\langle \varphi_k, \mathbb{1} \rangle (M(x) - \nu) + c \equiv 0, \quad c = \int_0^\omega N(t)\varphi_k(t)\,\mathrm{d}t,$$

which implies that

$$\langle \varphi_k, \mathbb{1} \rangle = 0, \quad \int_0^\omega N(t)\varphi_k(t)\,\mathrm{d}t = 0 \tag{4.1.27}$$

for the case $M(x) \not\equiv \text{const}$, where const stands for a constant function. Moreover, in the case $M(x) \equiv \text{const}$ the eigensubspace of S is infinite dimensional, and so the inequality $\dim H_\nu < \infty$ yields $M(x) \not\equiv \text{const}$. Thus, (4.1.25) holds.

Taking into account the second equalities in (4.1.25), we (similar to Remark 4.1.2) see that $H_\nu \in \mathcal{H}_+$ in the case $\langle \varphi_r, \mathbb{1} \rangle \ne 0$. By virtue of the first equalities in (4.1.25) it is immediate that $\langle \varphi_r, \mathbb{1} \rangle = 0$ yields $H_\nu \in \mathcal{H}_-$. □

From (4.1.25) and Lemma 4.1.3 we obtain that

$$\int_0^\omega \varphi_r(x)\,\mathrm{d}x \ne 0 \quad \text{when } H_\nu \in \mathcal{H}_+, \tag{4.1.28}$$

$$\int_0^\omega \varphi_r(x)N(x)\,\mathrm{d}x \ne 0 \quad \text{when } H_\nu \in \mathcal{H}_-. \tag{4.1.29}$$

4.2   On the distribution of the roots of Fourier images

1. In this section, we assume that the bounded operator S of the form (4.0.1) acts in $L^2(0, \omega)$ and is self-adjoint. Then the following equality is valid:

$$M(x) = \overline{N(x)}. \tag{4.2.1}$$

Remark 4.2.1. In view of (1.1.43) we have $S = USU$. Hence, the equality $Sf = \nu f$ yields $SUf = \nu Uf$, which, in turn, yields $S(f + Uf) = \nu(f + Uf)$ Therefore, for each nontrivial H_ν there is such a function $\varphi \in H_\nu$ ($\varphi \neq 0$) that $\varphi = \varphi^*$. More precisely, we choose $\varphi = cf + U(cf)$ ($c \in \mathbb{C}$). If there is a function $\varphi \in H_\nu$ such that $\langle \varphi, \mathbb{1} \rangle \neq 0$, then we can also choose some function $\varphi \in H_\nu$, which has both properties:

$$\langle \varphi, \mathbb{1} \rangle \neq 0, \quad \varphi = \varphi^*. \tag{4.2.2}$$

If there is a function $\varphi \in H_\nu$ such that $\langle \varphi, \overline{N} \rangle \neq 0$, then we can also choose some function $\varphi \in H_\nu$, which has both properties:

$$\langle \varphi, \overline{N} \rangle \neq 0, \quad \varphi = \varphi^*. \tag{4.2.3}$$

According to Proposition 4.1.6 and Remark 4.2.1, for $H_\nu \in \mathcal{H}_+$, there is $\varphi \in H_\nu$, which satisfies (4.2.2). Using (4.2.2), we rewrite (4.1.11) in the form:

$$\Phi(x) = \Psi(x) = \int_x^\omega \frac{\varphi(t)\, dt}{\langle \varphi, \mathbb{1} \rangle}. \tag{4.2.4}$$

Then $B = C$ and the equality (4.1.19) takes the form

$$B(S - \nu I) - (S - \nu I)B^* = 0. \tag{4.2.5}$$

If $H_\nu \subset \mathcal{H}_-$, we set

$$\Phi(x) = \Psi(x) = \int_x^\omega \frac{\varphi(t)\, dt}{\langle \varphi, \overline{N} \rangle}. \tag{4.2.6}$$

We note that the functions Φ given in (4.2.6) and in (4.1.21) coincide but Ψ in (4.2.6) differs from Ψ in (4.1.21). Thus, the first equality in (4.1.22) is valid for the case of $H_\nu \subset \mathcal{H}_-$ and Φ given by (4.2.6). Since $S = S^*$, it is immediate that $\nu = \overline{\nu}$, and so the second equality in (4.1.22) coincides with the first one. Relations (4.1.22) yield (4.1.19) for the case of P and Q given by (4.1.23). Moreover, from (4.1.23), (4.2.1) and (4.2.6) we again obtain $P = Q$, and so $B = C$. Taking into account $B = C$, we rewrite (4.1.19) in the form (4.2.5).

2. For $H_\nu \subset \mathcal{H}_+$, we set

$$G(\lambda) = \langle (A - \lambda I)^{-1} M, \Phi \rangle, \tag{4.2.7}$$

$$F(\lambda) = 1 - i\langle (A - \lambda I)^{-1} \mathbb{1}, \Phi \rangle. \tag{4.2.8}$$

If $H_\nu \subset \mathcal{H}_-$, we introduce G and F in a different way:

$$G(\lambda) = \langle (A - \lambda I)^{-1} \mathbb{1}, \Phi \rangle, \tag{4.2.9}$$

$$F(\lambda) = 1 - i\left\langle (A - \lambda I)^{-1} \left(M(x) - \frac{\nu}{2} \right), \Phi \right\rangle. \tag{4.2.10}$$

According to (4.1.15) and (4.1.16) we have $B^* f = A^* f + i\langle f, \mathbb{1} \rangle \Phi$, and so

$$(B^* - \lambda I)f = (A^* - \lambda I)f + i\langle f, \mathbb{1} \rangle \Phi = (A^* - \lambda I)\left(f + i\langle f, \mathbb{1} \rangle (A^* - \lambda I)^{-1} \Phi \right).$$

Hence, setting $f = (A^* - \lambda I)^{-1} \Phi$ we derive

$$(B^* - \lambda I)(A^* - \lambda I)^{-1} \Phi = \left(1 + i\langle (A^* - \lambda I)^{-1} \Phi, \mathbb{1} \rangle \right) \Phi,$$

that is,

$$(B^* - \lambda I)^{-1} \Phi = \frac{(A^* - \lambda I)^{-1} \Phi}{F_1(\lambda)}, \qquad H_\nu \in \mathcal{H}_+, \tag{4.2.11}$$

$$F_1(\lambda) = 1 + i\langle (A^* - \lambda I)^{-1} \Phi, \mathbb{1} \rangle = \overline{F(\overline{\lambda})}. \tag{4.2.12}$$

In a similar way we obtain:

$$(B - \lambda I)^{-1} \left(M(x) - \frac{\nu}{2} \right) = \frac{(A - \lambda I)^{-1} \left(M(x) - \frac{\nu}{2} \right)}{F(\lambda)}, \qquad H_\nu \in \mathcal{H}_-, \tag{4.2.13}$$

$$(B - \lambda I)^{-1} \mathbb{1} = \frac{(A - \lambda I)^{-1} \mathbb{1}}{F(\lambda)}, \qquad H_\nu \in \mathcal{H}_-. \tag{4.2.14}$$

From (4.2.10) we have

$$\left\langle (B - \lambda I)^{-1} g, \Phi \right\rangle = \frac{\left\langle (A - \lambda I)^{-1} g, \Phi \right\rangle}{F(\lambda)}. \tag{4.2.15}$$

We set $g(x) = M(x) + \beta_\nu$ in the case $H_\nu \in \mathcal{H}_+$ and $g(x) = \mathbb{1}$ in the case $H_\nu \in \mathcal{H}_-$. Then, from (4.1.14), (4.2.11) in the case $H_\nu \in \mathcal{H}_+$ and from (4.1.22) and (4.2.15) in the case $H_\nu \in \mathcal{H}_-$, we deduce

$$\frac{G(\lambda)}{F(\lambda)} = \langle (S - \nu I) (B^* - \lambda I)^{-1} \Phi, \Phi \rangle. \tag{4.2.16}$$

Rewrite (4.2.5) in the form

$$(S - \nu I) (B^* - \lambda I)^{-1} = (B - \lambda I)^{-1}(S - \nu I).$$

Then, from (4.2.16) it follows that

$$\frac{F(\lambda)}{G(\lambda)} = \overline{F(\overline{\lambda})}\,\Big/\,\overline{G(\overline{\lambda})}. \tag{4.2.17}$$

Formula (4.2.5) also yields the following identity

$$(S - \nu I)\,(B^* - \overline{\mu}I)^{-1} - (B - \lambda I)^{-1}(S - \nu I)$$
$$= (\overline{\mu} - \lambda)\,(B - \lambda I)^{-1}(S - \nu I)\,(B^* - \overline{\mu}I)^{-1}.$$

This implies that the relation

$$\big\langle (S - \nu I)\,(B^* - \overline{\mu}I)^{-1}\Phi, \Phi \big\rangle - \big\langle (S - \nu I)\Phi, (B^* - \overline{\lambda}I)^{-1}\Phi \big\rangle$$
$$= (\overline{\mu} - \lambda)\,\big\langle (S - \nu I)\,(B^* - \overline{\mu}I)^{-1}\Phi, (B^* - \overline{\lambda}I)^{-1}\Phi \big\rangle \tag{4.2.18}$$

is valid. From (4.2.11), (4.2.16) and (4.2.18) we deduce a basic for our study formula:

$$\frac{G(\lambda)\overline{F(\mu)} - \overline{G(\mu)}F(\lambda)}{\lambda - \overline{\mu}} = \big\langle (S - \nu I)\,(A^* - \overline{\mu}I)^{-1}\Phi, (A^* - \overline{\lambda}I)^{-1}\Phi \big\rangle. \tag{4.2.19}$$

3. Below we assume that the nonpositive spectrum of $S - \nu I$ consists of eigenvalues and denote the number of the negative eigenvalues of the operator $S - \nu I$ (counted with their multiplicities) by κ. By r we denote the dimension of H_ν.

By λ_j and μ_j (Im $\lambda_j \geq 0$, Im $\mu_j \geq 0$) we denote the roots in the upper half-plane of the functions $G(\lambda)$ and $F(\lambda)$, respectively. The values $K_G(\lambda_j)$ and $K_F(\mu_j)$ stand for the multiplicities of these roots.

According to Remark 4.2.1 the relation $f \in H_\nu$ yields $Uf \in H_\nu$. Recall that (for the case $r < \infty$) the functions φ_k of the form (4.1.24) form a basis in H_ν. In particular, the functions $U\varphi_k$ belong to H_ν. Moreover, in view of the first equalities in (4.1.25) we have $A^*\varphi_k = A\varphi_k = UA^*U\varphi_k$, that is,

$$UA^*\varphi_k = A^*U\varphi_k, \quad k < r. \tag{4.2.20}$$

Hence, taking into account (4.1.24), we see that, if $U\varphi_k \in ((A^*)^{k-1}H_\nu \cap H_\nu)$, then $U\varphi_{k+1} \in ((A^*)^k H_\nu \cap H_\nu)$. Thus, we show by induction that

$$U\varphi_r \in ((A^*)^{r-1}H_\nu \cap H_\nu).$$

We note that φ_r also belongs to $(A^*)^{r-1}H_\nu \cap H_\nu$ and that, according to (2.2.38), $\dim((A^*)^{r-1}H_\nu \cap H_\nu) = 1$. It is immediate that

$$\varphi_r = \theta U\varphi_r \quad (\theta \in \mathbb{C}, \quad |\theta| = 1). \tag{4.2.21}$$

Choosing $c \neq 0$ such that $\theta = \overline{c}/c$, we rewrite (4.2.21) in the form $c\varphi_r = U(c\varphi_r)$ (i.e., $(c\varphi_r) = (c\varphi_r)^*$). Therefore, in view of (4.1.28) and (4.1.29) we have the following corollary.

Corollary 4.2.2. *Without loss of generality we can choose φ_1 so that φ_r satisfies (4.2.2) for the case $H_\nu \in \mathcal{H}_+$ and satisfies (4.2.3) for the case $H_\nu \in \mathcal{H}_-$.*

Further we assume that φ_r is chosen in the way described in Corollary 4.2.2. Hence, the equality (4.2.19) is valid for the case of $H_\nu \in \mathcal{H}_+$ and Φ given by (4.2.4), as well as for the case of $H_\nu \in \mathcal{H}_-$ and Φ given by (4.2.6), where $\varphi = \varphi_r$. Recall that the notation κ, which we use in the following theorem, is introduced at the beginning of point 3.

Theorem 4.2.3. *Let S be a bounded self-adjoint operator of the form* (4.0.1) *acting on $L^2(0,\omega)$. If $\dim H_\nu = r$ $(1 \le r < \infty)$ and $\varphi(x) = \varphi_r(x)$, then we have*

$$\sum_{\operatorname{Im}\lambda_j > 0} K_G(\lambda_j) + \sum_{\operatorname{Im}\lambda_j = 0} \left\lfloor \frac{K_G(\lambda_j)}{2} \right\rfloor \le \kappa, \qquad (4.2.22)$$

$$\sum_{\operatorname{Im}\mu_j > 0} K_F(\mu_j) + \sum_{\operatorname{Im}\mu_j = 0} \left\lfloor \frac{K_F(\mu_j)}{2} \right\rfloor \le \kappa, \qquad (4.2.23)$$

where $\lfloor K \rfloor$ denotes the integer part of K and the functions $F(\lambda)$ and $G(\lambda)$ are given by the relations (4.2.4)–(4.2.10).

Proof. Differentiating both parts of (4.2.19), for the values p and q such that

$$1 \le p \le K_G(\lambda_j), \quad 1 \le q \le K_G(\lambda_l), \quad \lambda_j \ne \lambda_l, \qquad (4.2.24)$$

we obtain

$$\left\langle (S - \nu I)\,(A^* - \overline{\lambda}_j I)^{-p}\Phi,\, (A^* - \overline{\lambda}_l I)^{-q}\Phi \right\rangle = 0. \qquad (4.2.25)$$

Next, we deduce an analogue of (4.2.25) for the case $j = l$, $\lambda_j = \overline{\lambda}_j$. First, we multiply both parts of (4.2.19) by $(\lambda - \overline{\mu})$ and differentiate the obtained equality p times with respect to λ. In this way, for $\lambda = \overline{\lambda}$ and $\mu = \overline{\mu}$ we derive

$$G^{(p)}(\lambda)\overline{F(\mu)} - \overline{G(\mu)}F^{(p)}(\lambda)$$
$$= p!\Big((\lambda - \mu) \left\langle (S - \nu I)\,(A^* - \mu I)^{-1}\Phi,\, (A^* - \lambda I)^{-1-p}\Phi \right\rangle$$
$$+ \left\langle (S - \nu I)\,(A^* - \mu I)^{-1}\Phi,\, (A^* - \lambda I)^{-p}\Phi \right\rangle \Big), \quad 1 \le p. \qquad (4.2.26)$$

Differentiating (4.2.26) with respect to μ, we deduce

$$G^{(p)}(\lambda)\overline{F^{(q)}(\mu)} - \overline{G^{(q)}(\mu)}F^{(p)}(\lambda)$$
$$= p!\,q!\Big((\lambda - \mu) \left\langle (S - \nu I)\,(A^* - \mu I)^{-q-1}\Phi,\, (A^* - \lambda I)^{-1-p}\Phi \right\rangle$$
$$- \left\langle (S - \nu I)\,(A^* - \mu I)^{-q}\Phi,\, (A^* - \lambda I)^{-1-p}\Phi \right\rangle$$
$$+ \left\langle (S - \nu I)\,(A^* - \mu I)^{-1-q}\Phi,\, (A^* - \lambda I)^{-p}\Phi \right\rangle \Big). \qquad (4.2.27)$$

Setting
$$\mu = \lambda = \lambda_j = \overline{\lambda}_j, \quad 1 \leq p, q \leq K_G(\lambda_j) - 1, \tag{4.2.28}$$

from (4.2.26) and (4.2.27) we obtain

$$\left\langle (S - \nu I)(A^* - \lambda_j I)^{-1}\Phi, (A^* - \lambda_j I)^{-p}\Phi \right\rangle = 0, \tag{4.2.29}$$

$$-\left\langle (S - \nu I)(A^* - \lambda_j I)^{-q}\Phi, (A^* - \lambda_j I)^{-p-1}\Phi \right\rangle$$
$$+ \left\langle (S - \nu I)(A^* - \lambda_j I)^{-q-1}\Phi, (A^* - \lambda_j I)^{-p}\Phi \right\rangle = 0. \tag{4.2.30}$$

Now, let p and q be such that

$$p + q \leq K_G(\lambda_j), \quad p \geq 1, \quad q \geq 1. \tag{4.2.31}$$

Then, using (4.2.29) and (4.2.30) we calculate directly that

$$\left\langle (S - \nu I)(A^* - \lambda_j I)^{-q}\Phi, (A^* - \lambda_j I)^{-p}\Phi \right\rangle = 0. \tag{4.2.32}$$

We note that (4.2.32) also holds for $\lambda_j \neq \overline{\lambda}_j$.

By $\mathcal{L}_0$ we denote the space spanned on the functions

$$\Phi_{j,p} = (A^* - \overline{\lambda}_j I)^{-p}\Phi, \quad 1 \leq p \leq m_j, \tag{4.2.33}$$

where $m_j = K_G(\lambda_j)$ in the case $\mathrm{Im}\,\lambda_j > 0$, and $m_j = \lfloor K_G(\lambda_j)/2 \rfloor$ in the case $\mathrm{Im}\,\lambda_j = 0$. Let us prove that the system of the functions $\Phi_{j,p}$ is linearly independent. Indeed, from (4.2.33) we deduce that

$$\Phi_{j,p}(x) = (-\overline{\lambda}_j)^{-p}\Phi(x) + \int_x^\omega \Phi(t)e^{-(\mathrm{i}/\overline{\lambda}_j)(x-t)}Q_{p-1,j}(x-t)\,\mathrm{d}t, \tag{4.2.34}$$

where $Q_{p-1,j}$ is a polynomial of the order $(p-1)$. If the linear dependence relation

$$\sum_{j,p} \alpha_{j,p}\Phi_{j,p} = 0 \tag{4.2.35}$$

(where $\alpha_{j,p} \neq 0$) is fulfilled, from (4.2.34), (4.2.35) and the fact that the spectrum of Volterra operators is located at zero, we obtain

$$\int_x^\omega \Phi(t) \sum_{j,p} \alpha_{j,p}e^{-(\mathrm{i}/\overline{\lambda}_j)(x-t)}Q_{p-1,j}(x-t)\,\mathrm{d}t = 0.$$

Hence, by the Titchmarsh theorem (see [104, Chap. 11]) we have

$$\sum_{j,p} \alpha_{j,p}e^{-(\mathrm{i}x/\overline{\lambda}_j)}Q_{p-1,j}(x) = 0, \quad 0 \leq x \leq \beta,$$

for some $0 < \beta \leq \omega$. This means that $\alpha_{j,p} \equiv 0$, that is, we arrive at a contradiction. Therefore, the functions $\Phi_{j,p}$ are linearly independent and so the following relation is valid:

$$\dim \mathcal{L}_0 = \sum_j m_j. \tag{4.2.36}$$

Moreover, by virtue of (4.2.25) (for the case $\lambda_j \neq \lambda_l$) and (4.2.32), we have

$$\langle (S - \nu I)f, g \rangle = 0; \quad f, g \in \mathcal{L}_0. \tag{4.2.37}$$

Since $(S - \nu I)H_\nu = 0$, we rewrite (4.2.37) in the form

$$\langle (S - \nu I)f, g \rangle = 0; \quad f, g \in (\mathcal{L}_0 + H_\nu). \tag{4.2.38}$$

Recall that $\Phi(x) = \alpha A^{*r}\varphi_1$ ($\alpha \neq 0$). Therefore, it is easy to see that the set of functions $\{\varphi_k\} \cup \{\Phi_{j,p}\}$ is linearly independent. Indeed, if these functions are linearly dependent, formulas (4.1.24) and (4.2.33) imply that there are some integer $k < r$, constant $\alpha_k \neq 0$ and upper triangular operator T such that

$$(A^*)^k(\alpha_k I + A^*T)\varphi_1 = 0. \tag{4.2.39}$$

However, $(A^*)^k$ does not have eigenvectors, and so we can rewrite (4.2.39) in the form $(\alpha_k I + A^*T)\varphi_1 = 0$. Taking into account that $\alpha_k \neq 0$ and A^*T is a Volterra operator, we easily deduce from $(\alpha_k I + A^*T)\varphi_1 = 0$ that $\varphi_1 = 0$. Thus, we arrive at a contradiction, which means that our set of functions is linearly independent. Now, the relation

$$H_\nu \cap \mathcal{L}_0 = 0 \tag{4.2.40}$$

is immediate. From (4.2.38) we obtain $\dim(\mathcal{L}_0 + H_\nu) \leq \kappa + r$. Hence, using (4.2.40) we have $\dim \mathcal{L}_0 \leq \kappa$. Finally, (4.2.22) follows from the equality (4.2.36) and inequality $\dim \mathcal{L}_0 \leq \kappa$. Formula (4.2.23) is proved in a similar way. □

4. Introduce $\mathcal{D}_r(z)$ by the equality

$$\mathcal{D}_r(z) = \int_0^\omega e^{izx} \overline{\varphi_r(x)} \, dx. \tag{4.2.41}$$

Denote by z_j the roots of $\mathcal{D}_r(z)$ and by $K_{\mathcal{D}_r}(z_j)$ the multiplicities of these roots.

Theorem 4.2.4. *Let the conditions of Theorem 4.2.3 be fulfilled. Then the following inequality is valid:*

$$\sum_{\mathrm{Im}\, z_j < 0} K_{\mathcal{D}_r}(z_j) + \sum_{\mathrm{Im}\, z_j = 0, z_j \neq 0} \left\lfloor (K_{\mathcal{D}_r}(z_j)/2) \right\rfloor \leq \kappa. \tag{4.2.42}$$

Proof. If $H_\nu \in \mathcal{H}_+$, formula (4.2.8) takes the form

$$F(\lambda) = 1 + \frac{\mathrm{i}}{\lambda} \int\limits_0^\omega \mathrm{e}^{\mathrm{i}x/\lambda} \overline{\Phi(x)} \, \mathrm{d}x. \tag{4.2.43}$$

Substituting $\varphi = \varphi_r$ into (4.2.4), substituting the result into (4.2.43) and integrating by parts, we obtain

$$F(\lambda) = \alpha \mathcal{D}_r \left(\frac{1}{\lambda} \right), \quad \alpha \neq 0. \tag{4.2.44}$$

If $H_\nu \in \mathcal{H}_-$, then according to (4.2.9) and (4.2.41) we have

$$G(\lambda) = \alpha \mathcal{D}_r \left(\frac{1}{\lambda} \right), \quad \alpha \neq 0. \tag{4.2.45}$$

Now, the required inequality (4.2.42) follows from (4.2.44), (4.2.45) and Theorem 4.2.3.   □

A similar fact for the operators of the form (4.0.5) was proved by M.G. Kreĭn and was reported by him at a seminar in Odessa in the middle of the sixties (see [42]).

According to (4.1.24), any function φ from H_ν can be represented in the form

$$\varphi = \alpha_0 \varphi_r + \alpha_1 \varphi_r' + \cdots + \alpha_{r-1} \varphi_r^{(r-1)}. \tag{4.2.46}$$

From (4.1.2) and (4.1.24) for the case $H_\nu \in \mathcal{H}_-$ and from (4.1.24) and (4.1.25) for the case $H_\nu \in \mathcal{H}_+$, we easily derive

$$\varphi_r^{(k)}(0) = \varphi_r^{(k)}(\omega) = 0, \quad 0 \le k \le r - 2. \tag{4.2.47}$$

Now, we set

$$\mathcal{D}(z) = \int\limits_0^\omega \mathrm{e}^{\mathrm{i}zx} \overline{\varphi(x)} \, \mathrm{d}x. \tag{4.2.48}$$

Integrating (4.2.48) by parts and taking into account (4.2.47) we obtain

$$\mathcal{D}(z) = P_r(z) \mathcal{D}_r(z), \tag{4.2.49}$$

where

$$P_r(z) = \overline{\alpha}_0 + \overline{\alpha}_1(-\mathrm{i}z) + \cdots + \overline{\alpha}_r(-\mathrm{i}z)^{r-1}.$$

Example 4.2.5. Consider the operator S given by (4.0.6) and enumerate its eigenvalues in the order of their decrease. This operator has the following properties [46, 47, 63].

I. The operator S is non-negative.

II. All the eigenvalues of S are simple.

III. The equalities

$$\mu_n \varphi_n(x) = \int_{-1}^{1} e^{icxt} \varphi_n(t) \, dt, \quad n = 0, 1, 2, \ldots, \qquad (4.2.50)$$

where $\varphi_n(x)$ are the eigenfunctions corresponding to the eigenvalues

$$\lambda_n = \frac{c}{2\pi} |\mu_n|^2$$

of S, are valid.

IV. The function $\varphi_n(x)$ has n roots (counting their multiplicities) inside the segment $[-1, 1]$.

Using (4.2.50) we continue $\varphi_n(x)$ analytically on the complex plane

$$\varphi_n(z) = \frac{1}{\mu_n} \int_{-1}^{1} e^{iczt} \varphi_n(t) \, dt.$$

Applying Theorem 4.2.4 to the operator $(-S)$ we obtain the next proposition.

Proposition 4.2.6. *The multiplicities $K(z_j)$ of the roots of $\varphi_n(z)$ satisfy the inequality*

$$\sum_{\operatorname{Im} z_j < 0} K(z_j) + \sum_{\operatorname{Im} z_j = 0, z_j \neq 0} \left\lfloor \frac{K(z_j)}{2} \right\rfloor \leq n.$$

Chapter 5

Integral Operators with W-Difference Kernels

In Section 5.1 we consider a set of the operators A, S, B, Π_1, Γ_1, Π_2^*, Γ_2^* satisfying the relations

$$AS - SB = \Pi_1 \Pi_2^*, \tag{5.0.1}$$

$$S\Gamma_1 = \Pi_1, \quad \Gamma_2^* S = \Pi_2^*. \tag{5.0.2}$$

Such a set is called an S-node. The theory of S-nodes allows us to use one and the same approach in the study of various analysis problems: system theory [79], interpolation problems [27], inverse and direct spectral problems [90] and integrable equations [65, 91] (see also the books [69, 93, 94], which appeared after the first edition of this book was published, and numerous references therein). In this section we do not suppose that the spaces, in which the S-node operators act, are always Hilbert spaces.

We use the general results of Section 5.1 in further sections, where operators with a W-difference kernel (i.e., operators from $S(W)$ class) are investigated.

Operators with a difference kernel belong to the $S(W)$ class. The $S(W)$ class also contains the operators of the form

$$S_\Delta f = \mu f(x) + \int_\Delta k(x - t) f(t)\, \mathrm{d}t = \varphi(x), \quad x \in \Delta, \tag{5.0.3}$$

where Δ denotes a system of segments, which do not intersect. The operators S_Δ play an important role in elasticity theory (in the cases, where some elastic bodies interact in multiple connected domains) and in diffraction theory [25].

The Prandtl operator

$$\mathcal{A}f = -\frac{1}{\pi} \frac{\mathrm{d}}{\mathrm{d}x} \int_\Delta f'(t) \ln |x - t|\, \mathrm{d}t, \quad x \in \Delta \tag{5.0.4}$$

is also studied in this chapter. It turns out that some results of Chapters 1–3 obtained for the operators with difference kernels can be used for the operators of the class $S(W)$.

5.1   The principal notions of S-node theory

1. The class of the bounded linear operators acting from the Banach space H_1 into the Banach space H_2 is denoted by $\{H_1, H_2\}$. We assume that the operators from the S-node are bounded and, more precisely, that

$$A, B, S \in \{H, H\}, \quad \Pi_1, \Gamma_1 \in \{G, H\}, \quad \Pi_2^*, \Gamma_2^* \in \{H, G\}, \tag{5.1.1}$$

where G and H are some Banach spaces. Recall that the operators (5.1.1) form an S-node if the following relations

$$AS - SB = \Pi_1 \Pi_2^*, \tag{5.1.2}$$

$$S\Gamma_1 = \Pi_1, \quad \Gamma_2^* S = \Pi_2^* \tag{5.1.3}$$

are fulfilled. The transfer operator functions

$$w_A(z) = I - \Gamma_2^*(A - zI)^{-1}\Pi_1, \quad w_B(z) = I + \Pi_2^*(B - zI)^{-1}\Gamma_1 \tag{5.1.4}$$

play an essential role in system theory. We shall need the following statement from [79].

Theorem 5.1.1. *Assume that the values z and ζ do not belong to the spectrum of the operators A and B, respectively. Then the equality*

$$w_A(z)w_B(\zeta) = I - (z - \zeta)\Gamma_2^*(A - zI)^{-1}S(B - \zeta I)^{-1}\Gamma_1 \tag{5.1.5}$$

is valid.

Proof. Indeed,

$$\begin{aligned} w_A(z)w_B(\zeta) = I &- \Gamma_2^*(A - zI)^{-1}\Pi_1 + \Pi_2^*(B - \zeta I)^{-1}\Gamma_1 \\ &- \Gamma_2^*(A - zI)^{-1}\Pi_1\Pi_2^*(B - \zeta I)^{-1}\Gamma_1. \end{aligned} \tag{5.1.6}$$

We write the equality (5.1.2) in the form

$$\Pi_1 \Pi_2^* = (A - zI)S - S(B - \zeta I) + (z - \zeta)S. \tag{5.1.7}$$

From (5.1.3), (5.1.6) and (5.1.7) we obtain the statement of the theorem (i.e., the equality (5.1.5)). $\qquad\square$

For the case $z = \zeta$, it follows from (5.1.5) that

$$w_A(z)w_B(z) = I. \tag{5.1.8}$$

The notation $\mathcal{D}(A, B)$ stands for a connected set of complex numbers, which contains $z = \infty$ but contains neither points of spectrum of the operator A nor of B. In the neighborhood of $z = \infty$, the operator function $w_A(z)$ is invertible. In view of (5.1.8), it means that

$$w_B(z) = w_A(z)^{-1}, \quad z \in \mathcal{D}(A, B). \tag{5.1.9}$$

Corollary 5.1.2. *For any $z \in \mathcal{D}(A, B)$, the equation*

$$SY(z) = (A - zI)^{-1}\Pi_1 \tag{5.1.10}$$

has a solution

$$Y(z) = (B - zI)^{-1}\Gamma_1 w_A(z). \tag{5.1.11}$$

Proof. Indeed, according to (5.1.2) the relation

$$S(B - zI)^{-1} - (A - zI)^{-1}S = (A - zI)^{-1}\Pi_1\Pi_2^*(B - zI)^{-1}$$

is valid. Hence, from (5.1.3) and (5.1.5) we obtain

$$S(B - zI)^{-1}\Gamma_1 = (A - zI)^{-1}\Pi_1 w_B(z). \tag{5.1.12}$$

From (5.1.9) and (5.1.12) we deduce the statement of the corollary. □

2. Let us consider in detail the important subcase where

$$G = G_1 \oplus G_1 \tag{5.1.13}$$

and G_1 is a Hilbert space. In this case

$$g = \begin{bmatrix} g_1 \\ g_2 \end{bmatrix}, \quad \Pi_1 = \begin{bmatrix} \Phi_1 & \Phi_2 \end{bmatrix}, \quad \Pi_2^* = \begin{bmatrix} \Psi_1^* \\ \Psi_2^* \end{bmatrix}, \tag{5.1.14}$$

where $g_k \in G_1$; $\Phi_1, \Phi_2 \in \{G_1, H\}$; $\Psi_1^*, \Psi_2^* \in \{H, G_1\}$. The relation (5.1.2) takes the form

$$AS - SB = \Phi_1\Psi_1^* + \Phi_2\Psi_2^*. \tag{5.1.15}$$

Setting

$$\Gamma_1 = \begin{bmatrix} \widetilde{\Phi}_1 & \widetilde{\Phi}_2 \end{bmatrix}, \quad \Gamma_1^* = \begin{bmatrix} \widetilde{\Psi}_1^* \\ \widetilde{\Psi}_2^* \end{bmatrix}, \tag{5.1.16}$$

we rewrite (5.1.3) in the form

$$S\widetilde{\Phi}_k = \Phi_k, \ \widetilde{\Phi}_k \in \{G_1, H\}, \quad \widetilde{\Psi}_k^*S = \Psi_k, \ \widetilde{\Psi}_k^* \in \{H, G_1\}, \quad k = 1, 2. \tag{5.1.17}$$

We partition the operator functions $w_A(z)$ and $w_B(z)$ into the blocks in accordance with the decomposition (5.1.13):

$$w_A(z) = \begin{bmatrix} a_1(z) & a_2(z) \\ b_1(z) & b_2(z) \end{bmatrix}, \quad w_B(z) = \begin{bmatrix} c_1(z) & c_2(z) \\ d_1(z) & d_2(z) \end{bmatrix}. \tag{5.1.18}$$

From Corollary 5.1.2 it follows that the equations

$$SY_k(z) = (A - zI)^{-1}\Phi_k, \quad k = 1, 2, \tag{5.1.19}$$

have the following solutions:

$$Y_k(z) = (B - zI)^{-1}\left(\widetilde{\Phi}_1 a_k(z) + \widetilde{\Phi}_2 b_k(z)\right). \tag{5.1.20}$$

Now, we introduce the operator function

$$R(z, \zeta) = \Psi_1^*(B - zI)^{-1}Y_2(\zeta). \tag{5.1.21}$$

The reflection and transmission coefficients in the problems of radiation transfer [29, 100] and the scattering diagram in the problems of diffraction [24] are expressed in terms of $R(z, \zeta)$. The analytical structure of $R(z, \zeta)$ is characterized by the following theorem.

Theorem 5.1.3. *Let the relations* (5.1.14)–(5.1.16) *hold and assume that* $z, \zeta \in \mathcal{D}(A, B)$. *Then*

$$R(z, \zeta) = \frac{c_1(z)a_2(\zeta) + c_2(z)b_2(\zeta)}{z - \zeta}, \tag{5.1.22}$$

$$c_1(\zeta)a_2(\zeta) + c_2(\zeta)b_2(\zeta) = 0. \tag{5.1.23}$$

Proof. By virtue of (5.1.20) and (5.1.21) we have

$$R(z, \zeta) = \Psi_1^*\left((B - zI)^{-1} - (B - \zeta I)^{-1}\right)\frac{\widetilde{\Phi}_1 a_2(\zeta) + \widetilde{\Phi}_2 b_2(\zeta)}{z - \zeta}.$$

Hence, according to the last equalities in (5.1.4), (5.1.14) and (5.1.18) and the first equality in (5.1.16), we deduce

$$R(z, \zeta) = \frac{\left(c_1(z) - c_1(\zeta)\right)a_2(\zeta) + \left(c_2(z) - c_2(\zeta)\right)b_2(\zeta)}{z - \zeta}. \tag{5.1.24}$$

In view of (5.1.9) the equality (5.1.23) is valid. Finally, (5.1.22) is immediate from (5.1.23) and (5.1.24).   □

5.2 Operators with W-difference kernels

1. In this section, we consider the bounded operators

$$Sf = \frac{\mathrm{d}}{\mathrm{d}x} \int\limits_0^\omega s(x,t) f(t)\, \mathrm{d}t = \varphi(x), \quad S \in \{L_n^p(0,\omega), L_n^p(0,\omega)\}, \tag{5.2.1}$$

where $L_n^p(0,\omega)$ is the space of vector functions $f(x) = \begin{bmatrix} f_1(x) \\ \dots \\ f_n(x) \end{bmatrix}$ and the matrix

kernel $s(x,t)$ has the structure

$$s(x,t) = \big\{ s_{l,m}(\omega_l x - \omega_m t) \big\}_{l,m=1}^n. \tag{5.2.2}$$

Here, we suppose that

$$s_{l,m}(x) \in L^q(-\omega_m \omega, \omega_l \omega), \quad \frac{1}{p} + \frac{1}{q}, \quad \omega_m > 0 \quad (1 \le m \le n).$$

We denote by W the diagonal matrix

$$W = \mathrm{diag}\big\{ \omega_1, \omega_2, \dots, \omega_n \big\}.$$

Operators of the form (5.2.1), (5.2.2) are called the operators with W-difference kernels, whereas kernels of the form (5.2.2) are called the W-difference kernels. If $W = I_n$, the corresponding operator S is an operator with a difference kernel. In the contact theory of elasticity and in the problems of diffraction on a system of ribbons, an essential role is played by the equations of the form

$$Sf = \mu f(x) + \int\limits_\Delta k(x-t) f(t)\, \mathrm{d}t = \varphi(x), \quad x \in \Delta, \tag{5.2.3}$$

where Δ denotes a system of non-intersecting segments $[a_k, b_k]$ such that

$$a_1 < b_1 < a_2 < b_2 < \cdots < a_n < b_n.$$

The summarized length of the segments is denoted by mes, that is,

$$\mathrm{mes}\,\Delta = \sum_{k=1}^n \big(b_k - a_k \big). \tag{5.2.4}$$

The operator S can be transformed into the operator (5.2.1) with a W-difference kernel. For this we put

$$f(v) = \begin{bmatrix} f(a_1 + v\omega_1)\sqrt{\omega_1} \\ f(a_2 + v\omega_2)\sqrt{\omega_2} \\ \dots \\ f(a_n + v\omega_n)\sqrt{\omega_n} \end{bmatrix}, \quad \varphi(v) = \begin{bmatrix} \varphi(a_1 + v\omega_1)\sqrt{\omega_1} \\ \varphi(a_2 + v\omega_2)\sqrt{\omega_2} \\ \dots \\ \varphi(a_n + v\omega_n)\sqrt{\omega_n} \end{bmatrix},$$

where $\omega_k = b_k - a_k$. Then equation (5.2.3) takes the form

$$\mu f(v) + \int_0^1 K(v, u) f(u)\, du = \varphi(v), \tag{5.2.5}$$

where the matrix kernel $K(v, u)$ is defined by the formula

$$K(v, u) = \left\{ k(a_l - a_m + \omega_l v - \omega_m u) \sqrt{\omega_l \omega_m} \right\}_{l,m=1}^n. \tag{5.2.6}$$

It is easy to see that system (5.2.5), (5.2.6) is a particular case of the system (5.2.1), (5.2.2).

2. Using direct calculation we obtain the operator identity

$$\left(AS - SA^*\right) f = i \int_0^\omega \left(M(x) + N(t) \right) f(t)\, dt, \tag{5.2.7}$$

where $M(x) = W s(x, 0)$, $N(x) = -W s(0, x)$ and

$$A f = iW \int_0^x f(t)\, dt, \quad A^* f = -iW \int_x^\omega f(t)\, dt. \tag{5.2.8}$$

The identity (5.2.7) can be written down in the form (5.1.15). Indeed, let G_1 be the space $\mathbb{C}^n$ of the column vectors of order n. Introduce the operators

$$\Phi_1 g = M(x) g, \quad \Phi_2 g = g; \quad \Phi_1, \Phi_2 \in \{G_1, L_n^p(0, \omega)\}, \tag{5.2.9}$$

$$\Psi_1^* f(x) = i \int_0^\omega f(x)\, dx, \quad \Psi_2^* f(x) = i \int_0^\omega N(x) f(x)\, dx; \quad \Psi_1^*, \Psi_2^* \in \{L_n^p(0, \omega), G_1\}. \tag{5.2.10}$$

For the operators A and $B = A^*$ given by the equalities (5.2.8) and for the operators Φ_1, Φ_2, Ψ_1^*, Ψ_2^* given by the formulas (5.2.9), (5.2.10), the operator identity (5.2.7) takes the form (5.1.15). We also suppose that there are $n \times n$ matrix functions $N_k(x)$ and $M_k(x)$ $(k = 1, 2)$ with the entries from $L^p(0, \omega)$ satisfying the relations

$$SN_1 = M(x), \quad SN_2 = I_n, \tag{5.2.11}$$

$$\langle Sf, M_1 \rangle = \langle f, I_n \rangle, \quad \langle Sf, M_2 \rangle = \langle f, N^* \rangle, \quad f \in L_n^p(0, \omega). \tag{5.2.12}$$

The operator S acts in (5.2.11) columnwise, that is, S transforms columns of N_1 and N_2 into the corresponding columns of M and I_n, respectively. In (5.2.12) we use the notation

$$\langle C, D \rangle = \int_0^\omega D^*(x) C(x)\, dx, \tag{5.2.13}$$

where C and D are matrix functons. According to (5.2.11) and (5.2.12), the operators $\widetilde{\Phi}_k$ and $\widetilde{\Psi}_k$ given by (5.1.17) have the form

$$\widetilde{\Phi}_k g = N_k(x)g, \quad \widetilde{\Psi}_k^* f = i \int_0^\omega M_k^*(x)f(x)\,dx. \tag{5.2.14}$$

We apply the results of Section 5.1 in order to solve the equation

$$SB(x,\lambda) = e^{i\lambda x W}, \tag{5.2.15}$$

where the operator S has a W-difference kernel. For this purpose we introduce the matrices

$$a(\lambda) = i\lambda \int_0^\omega M_1^*(t)\,e^{i\lambda W t}\,dt, \tag{5.2.16}$$

$$b(\lambda) = I_n + i\lambda \int_0^\omega M_2^*(t)\,e^{i\lambda W t}\,dt. \tag{5.2.17}$$

Theorem 5.2.1. *Let an operator S with a W-difference kernel be bounded in the space $L_n^p(0,\omega)$ ($1 \le p \le 2$) and let some matrix functions $N_k(x)$ and $M_k(x)$ with entries belonging to $L^p(0,\omega)$ satisfy (5.2.11) and (5.2.12), respectively. Then the equation (5.2.15) has a solution*

$$B(x,\lambda) = u(x,\lambda) - i\lambda W \int_x^\omega e^{i\lambda(x-t)W} u(t,\lambda)\,dt, \tag{5.2.18}$$

where

$$u(x,\lambda) = N_1(x)a(\lambda) + N_2(x)b(\lambda). \tag{5.2.19}$$

Proof. Since

$$(I - \lambda A)^{-1} I_n = e^{i\lambda W x} \tag{5.2.20}$$

the equation (5.2.15) can be written in the form

$$SB(x,\lambda) = (I - \lambda A)^{-1}\Phi_2. \tag{5.2.21}$$

Using (5.1.20) we deduce that

$$B(x,\lambda) = (I - \lambda A^*)^{-1}\Big(N_1(x)a_2(1/\lambda) + N_2(x)b_2(1/\lambda)\Big). \tag{5.2.22}$$

Comparing (5.1.4), (5.1.18) and (5.2.16), (5.2.17) we have

$$a_2(1/\lambda) = a(\lambda), \quad b_2(1/\lambda) = b(\lambda). \tag{5.2.23}$$

Now, the statement of the theorem follows from (5.2.8) and (5.2.21)–(5.2.23). □

3. Let us introduce now the matrix functions

$$\rho(\lambda,\mu) = \int_0^\omega e^{ix\mu W} B(x,\lambda)\,dx; \tag{5.2.24}$$

$$c(\lambda) = I_n + i\lambda \int_0^\omega e^{i\lambda W x} N_1(x)\,dx, \quad d(\lambda) = i\lambda \int_0^\omega e^{i\lambda W x} N_2(x)\,dx. \tag{5.2.25}$$

Recall that $a(\lambda)$ and $b(\lambda)$ are defined by the relations (5.2.16) and (5.2.17).

The analytical structure of $\rho(\lambda,\mu)$ is characterized by the following theorem.

Theorem 5.2.2. *Let the conditions of Theorem 5.2.1 be fulfilled. Then the following equalities are valid:*

$$\rho(\lambda,\mu) = -i\,\frac{c(\mu)\,a(\lambda) + d(\mu)\,b(\lambda)}{\lambda+\mu}, \tag{5.2.26}$$

$$c(-\lambda)\,a(\lambda) + d(-\lambda)\,b(\lambda) = 0. \tag{5.2.27}$$

Proof. From Theorem 5.2.1 and formula (5.2.24) we deduce

$$\rho(\lambda,\mu) = -i\,\frac{\big(c(\mu) - c(-\lambda)\big)a(\lambda) + \big(d(\mu) - d(-\lambda)\big)b(\lambda)}{\lambda+\mu}. \tag{5.2.28}$$

Comparing (5.1.4), (5.1.18) and (5.2.25) we have

$$c_1(z) = c(-1/z), \quad d_1(z) = d(-1/z). \tag{5.2.29}$$

According to (5.1.23), (5.2.23) and (5.2.29), the equality (5.2.27) holds. The statement of the theorem follows from (5.2.27) and (5.2.28). $\qquad\square$

Formulas (5.1.22) and (5.2.26) are generalizations of the famous Ambartsumian formula [2], which was introduced for the case of operators S of the form

$$Sf = f(x) + \int_0^\omega k(x-t)f(t)\,dt, \tag{5.2.30}$$

where $k(x) = \overline{k(-x)} \in L(0,\omega)$. Recall that in Chapter 2 we modified Ambartsumian formula for the case of operators with a difference kernel acting in $L^p(0,\omega)$. The operator S^*, adjoint to S, acts in the space $L_n^q(0,\omega)$, $\frac{1}{p} + \frac{1}{q} = 1$, which is adjoint to $L_n^p(0,\omega)$. This operator is defined by the formula

$$\langle Sf, g\rangle = \langle f, S^*g\rangle, \tag{5.2.31}$$

where $f \in L_n^q(0,\omega)$, $g \in L_n^q(0,\omega)$ and the scalar product is given by

$$\langle f, g \rangle = \int_0^\omega g^*(x)f(x)\,\mathrm{d}x. \tag{5.2.32}$$

By virtue of (5.2.1) and (5.2.2) the operator S^* admits a representation

$$S^*f = \frac{\mathrm{d}}{\mathrm{d}x}\int_0^\omega s_1(x-t)f(t)\,\mathrm{d}t, \tag{5.2.33}$$

where

$$s_1(x,t) = -W^{-1}s^*(t,x)W. \tag{5.2.34}$$

We assume that the right-hand side of (5.2.33) is well defined in $L_n^p(0,\omega)$, and so S^* admits the corresponding extension onto this space. We note that (5.2.31) does not necessarily hold for this extension. Introduce the following $2n \times 2n$ matrix

$$\gamma = \left\{ \langle SN_k, M_l \rangle - \langle N_k, S^*M_l \rangle \right\}_{k,l=1}^2. \tag{5.2.35}$$

Problem 5.2.3. *Assuming that the conditions of Theorem 5.2.1 are fulfilled, describe the connection between the rank of γ and the dimension of the zero subspace of the operator S. (For the scalar case $n = 1$ the corresponding result is contained in Theorem 2.2.4.)*

5.3   The Prandtl equation

1. Recall that by Δ we denote a system of non-intersecting segments $[a_k, b_k]$ $(1 \leq k \leq m)$ where $a_1 < b_1 < a_2 < b_2 < \cdots < a_m < b_m$. We introduce the operators

$$Sf = -\frac{1}{\pi}\int_\Delta (\ln|x-t|)f(t)\,\mathrm{d}t, \quad x \in \Delta, \tag{5.3.1}$$

$$\mathcal{A}f = -\frac{\mathrm{d}}{\mathrm{d}x}S\frac{\mathrm{d}}{\mathrm{d}x}f = \varphi(x), \quad x \in \Delta. \tag{5.3.2}$$

The domain of definition $\mathcal{D}_\mathcal{A}$ of the operator $\mathcal{A}$ consists of absolutely continuous functions $f(x)$ such that

$$\frac{\mathrm{d}}{\mathrm{d}x}Sf' \in L^p(\Delta), \quad f(a_k) = f(b_k) = 0, \quad 1 < p < \infty, \quad 1 \leq k \leq m. \tag{5.3.3}$$

Equation (5.3.2) with the conditions (5.3.3) is called the Prandtl equation, because Prandtl used it in order to estimate the lifting capacity of a wing. This equation also plays an essential role in contact problems [20, 102], diffraction theory [24] and in the theory of stable processes [30].

We set

$$R(z) = \left(\prod_{k=1}^{m} (z - a_k)(z - b_k) \right)^{1/2}, \tag{5.3.4}$$

and choose a branch of $R(z)$, for which

$$\lim \left(z^{-m} R(z) \right) = 1, \quad z \to \infty. \tag{5.3.5}$$

It is known [55] that the following statement is valid.

Proposition 5.3.1. *Let $\varphi(x) \in L^p(\Delta)$ $(1 < p < \infty)$. Then the set of solutions of system (5.3.1), (5.3.2) is described by the equality*

$$f'(x) = \frac{1}{\pi R_+(x)} \left(\int_{\Delta} \frac{R_+(t)}{t - x} \varphi(t) \, dt + \pi P_{m-1}(x) \right), \tag{5.3.6}$$

where $x \in \Delta$, $P_{m-1}(x) = c_0 + c_1 x + \cdots + c_{m-1} x^{m-1}$, c_k are arbitrary constants, $R_+(x) = \lim_{\eta \to +0} R(z)$ $(z = \xi + i\eta)$.

In view of the conditions $f(a_k) = 0$ we obtain

$$f(x) = \frac{1}{\pi} \int_{a_k}^{x} \frac{1}{R_+(t)} \left(\int_{\Delta} \frac{R_+(u)}{u - x} \varphi(u) \, du + \pi P_{m-1}(t) \right) dt, \quad a_k \le x \le b_k. \tag{5.3.7}$$

Taking into account the conditions $f(b_k) = 0$, from (5.3.7) we deduce

$$\sum_{l=0}^{m-1} c_l \int_{a_k}^{b_k} \frac{u^l}{R_+(u)} \, du = \varphi_k, \tag{5.3.8}$$

where

$$\varphi_k = \int_{\Delta} \varphi(u) r_k(u) R_k(u) \, du, \quad r_k(u) = -\frac{1}{\pi} \int_{a_k}^{b_k} \frac{dt}{R_+(t)(u - t)}. \tag{5.3.9}$$

Introduce a matrix, which plays an important role in the theory:

$$\Gamma = \{ \gamma_{l,k} \}_{l,k=1}^{m}, \quad \gamma_{l,k} = \int_{a_k}^{b_k} \frac{u^{l-1}}{R_+(u)} \, du. \tag{5.3.10}$$

Proposition 5.3.2. *The following inequality is valid:*

$$\det \Gamma \neq 0. \tag{5.3.11}$$

Proof. We assume that the equalities

$$\int_{a_k}^{b_k} \frac{Q(u)}{R_+(u)}\, du = 0, \quad 1 \leq k \leq m \tag{5.3.12}$$

hold for a polynomial

$$Q(u) = Q_1(u) + iQ_2(u) = \alpha_1 + \alpha_2 u + \cdots + \alpha_m u^{m-1},$$

where $Q_1(u)$ and $Q_2(u)$ are real-valued polynomials. Since $\arg R_+(u)$ is constant on each of the segments $[a_k, b_k]$, it follows from (5.3.12) that

$$\int_{a_k}^{b_k} \frac{Q_r(u)}{R_+(u)}\, du = 0, \quad 1 \leq k \leq m, \quad r = 1, 2. \tag{5.3.13}$$

According to (5.3.13), the polynomials $Q_r(u)$ $(r = 1, 2)$ have roots on each of the segments $[a_k, b_k]$ $(1 \leq k \leq m)$. Taking into account the inequalities

$$\deg Q_r(u) \leq m - 1 \quad (r = 1, 2),$$

we obtain $Q_1(u) = Q_2(u) = Q(u) \equiv 0$. Hence, the proposition is proved. $\square$

From relation (5.3.9), the equality

$$R_+(t) = -R_-(t), \quad t \in \Delta \tag{5.3.14}$$

and the Sokhotski–Plemelj formula, it follows that

$$\sum_{k=1}^{m} r_k(u) = 0, \quad u \in \Delta. \tag{5.3.15}$$

The function $r_k(u)$ is infinitely differentiable in the domain $\Delta \backslash [a_k, b_k]$. The infinite differentiability of $r_k(u)$ when $u \in \Delta$ follows from (5.3.15). Using the relations (5.3.8) and (5.3.9) we derive that

$$f(x) = B\varphi = \int_{\Delta} B(x, t)\varphi(t)\, dt, \quad x \in \Delta, \tag{5.3.16}$$

where the kernel $B(x,t)$ of the operator $\mathcal{B}$ is defined by the formulas

$$B(x,t) = B_1(x,t) + B_2(x,t), \tag{5.3.17}$$

$$B_1(x,t) = \frac{1}{\pi} \int\limits_{a_k}^{x} \frac{1}{R_+(u)\,(t-u)}\, du, \quad a_k \le x \le b_k, \quad 1 \le k \le m, \tag{5.3.18}$$

$$B_2(x,t) = R_+(t) \begin{bmatrix} r_1(t) & r_2(t) & \cdots & r_m(t) \end{bmatrix} \Gamma^{-1} \begin{bmatrix} q_1(x) \\ q_2(x) \\ \cdots \\ q_m(x) \end{bmatrix}, \tag{5.3.19}$$

$$q_l(x) = \int\limits_{a_k}^{x} \frac{u^{l-1}}{R_+(u)}\, du. \tag{5.3.20}$$

We note that according to Proposition 5.3.2, Γ is invertible, and so Γ^{-1} is well defined.

Theorem 5.3.3. *If $\varphi(x) \in L^r(\Delta)$ $(1 < p < r \le 2)$, then equations (5.3.1)–(5.3.3) have (in $\mathcal{D}_A$) one and only one solution $f(x) = \mathcal{B}\varphi$.*

For the case of one segment (i.e., for $m = 1$), the solution of (5.3.2) was constructed earlier in [30].

2. Let $\lambda_n(\mathcal{B})$ be the spectrum, enumerated in the decreasing order, of the bounded self-adjoint operator $\mathcal{B}$. Recall that the notation mes Δ is introduced in (5.2.4).

Theorem 5.3.4. *The following asymptotic equality is valid:*

$$\lambda_n(\mathcal{B}) = \frac{\text{mes}\,\Delta}{\pi n}\,(1 + o(1)), \quad n \to \infty. \tag{5.3.21}$$

For the case $m = 1$, this fact was already known [10, 36]. The general case easily follows from this special case if one uses formulas (5.3.16)–(5.3.20) and the smoothness of $r_k(u)$.

Corollary 5.3.5. *The spectrum μ_n of the Prandtl operator, enumerated in the increasing order, satisfies the asymptotic equality*

$$\mu_n(\mathcal{A}) = \frac{\pi n}{\text{mes}\,\Delta}\,(1 + o(1)). \tag{5.3.22}$$

Problem 5.3.6. *Find the second term in the asymptotic equality (5.3.22).*

Chapter 6

Problems of Communication Theory

6.1 Optimal prediction

1. A stochastic process $X(t)$ is called stationary in the wide sense if the mathematical expectation $E[X(t)] = m$ is constant and the correlation function depends only on the difference $(t - s)$:

$$k_x(t, s) = E\left[X(t)\overline{X(s)}\right] = k_x(t - s). \tag{6.1.1}$$

Here it is assumed that

$$E\left[|X(t)|^2\right] < \infty. \tag{6.1.2}$$

2. Next, we consider a device with a finite memory ω which transforms the input stochastic process $\widetilde{X}(t)$ into the output stochastic process $\widetilde{Y}(t)$ according to the following law:

$$\widetilde{Y}(t) = \alpha\widetilde{X}(t) + \beta\widetilde{X}(t - \omega) + \int_{t-\omega}^{t} \widetilde{X}(s)g(t - s)\,\mathrm{d}s, \tag{6.1.3}$$

where

$$g(x) \in L(0, \omega).$$

Wiener's book [112] is devoted to the problem of the synthesis of the optimal device in the case $\omega = \infty$. His results were extended to the case $\omega < \infty$ by Zadeh and Ragazzini in the paper [113], in which it is assumed that the stochastic process has the form

$$\widetilde{X}(t) = P(t) + X(t), \tag{6.1.4}$$

where $P(t)$ is a polynomial of degree n with known coefficients and $X(t)$ is continuous and stationary in the wide sense and has a given correlation function $k_x(t - s)$.

The problem of forward optimal prediction for a time interval τ can be formulated in the following way [101]. It is required to select the charateristics α, β, and $g(x)$ of the device so that the output process $\widetilde{Y}(t)$ is as close as possible to the process $\widetilde{X}(t + \tau)$, where $\tau > 0$; moreover, α, β, and $g(x)$ are assumed to be such that the corresponding device transforms $P(t)$ into $P(t + \tau)$, that is,

$$P(t + \tau) = \alpha P(t) + \beta P(t - \omega) + \int_{t-\omega}^{t} P(s)g(t - s)\,ds. \qquad (6.1.5)$$

We estimate the distance between $\widetilde{Y}(t)$ and $\widetilde{X}(t + \tau)$ in terms of ε defined by

$$\varepsilon^2 = E\left[\left|\widetilde{X}(t + \tau) - \widetilde{Y}(t)\right|^2\right]. \qquad (6.1.6)$$

3. Putting
$$G(u) = \alpha\delta(u) + \beta\delta(\omega - u) + g(u), \qquad (6.1.7)$$

we rewrite (6.1.5) in the form

$$P(t + \tau) = \int_{0}^{\omega} P(t - u)G(u)\,du. \qquad (6.1.8)$$

Using the formula
$$P(t - u) = \sum_{r=0}^{n} (-1)^r P^{(r)}(t)\frac{u^r}{r!},$$

we deduce from (6.1.8)

$$\int_{0}^{\omega} u^r G(u)\,du = (-1)^r \tau^r, \quad 0 \le r \le n. \qquad (6.1.9)$$

Then (6.1.3), (6.1.6), and (6.1.7) define the quadratic functional ε^2 on the space of generalized functions $G(u)$. The prediction problem reduces to the minimization of this functional, subject to (6.1.9). The following prediction result is valid [101].

Theorem 6.1.1. *Suppose that the generalized function $G_0(u)$ has the form*

$$G_0(u) = \alpha_0\delta(u) + \beta_0\delta(\omega - u) + g_0(u), \quad g_0(u) \in L(0, \omega), \qquad (6.1.10)$$

and satisfies the conditions

$$\int_{0}^{\omega} u^r G_0(u)\,du = (-1)^r \tau^r, \quad 0 \le r \le n. \qquad (6.1.11)$$

Then $G_0(u)$ is a solution of the prediction problem if and only if the equality

$$\int_0^\omega G_0(v)k_x(u-v)\,dv = k_x(u+\tau) + \sum_{r=0}^n \gamma_r u^r \qquad (6.1.12)$$

holds for some coefficients γ_r.

Proof. Suppose that (6.1.10)–(6.1.12) are satisfied. Taking (6.1.9) into account we obtain

$$\int_0^\omega \left(k_x(u+\tau) - \int_0^\omega k_x(u-v)G_0(v)\,dv\right)\left(\overline{G(u)} - \overline{G_0(u)}\right)du = 0, \qquad (6.1.13)$$

which is equivalent to

$$E\left[(X(t+\tau) - Y_0(t))\left(\overline{Y(t)} - \overline{Y_0(t)}\right)\right] = 0, \qquad (6.1.14)$$

where

$$Y(t) = \int_{t-\omega}^t G(t-s)X(s)\,ds, \quad Y_0(t) = \int_{t-\omega}^t G_0(t-s)X(s)\,ds. \qquad (6.1.15)$$

By (6.1.8), the formula (6.1.6) takes the form

$$\varepsilon^2 = E\left[|X(t+\tau) - Y(t)|^2\right]. \qquad (6.1.16)$$

Using (6.1.14) we arrive at the equation

$$\varepsilon^2 = E\left[|X(t+\tau) - Y_0(t)|^2\right] + E\left[|Y(t) - Y_0(t)|^2\right]. \qquad (6.1.17)$$

If $G(u) = G_0(u)$, then $Y(t) = Y_0(t)$, and, by virtue of (6.1.17), the functional ε^2 attains its minimal value.

Now, let $G(u)$ be some function satisfying (6.1.10) and (6.1.11), and minimizing ε^2. The necessary condition for an extremum implies (6.1.14), and so (6.1.13) and (6.1.12) follow. This proves the theorem. $\qquad \square$

4. Let $\mathcal{F}_\tau(u)$ and $\mathcal{L}_r(u)$ be functions defined by

$$\int_0^\omega k_x(u-v)\mathcal{F}_\tau(v)\,dv = k_x(u+\tau), \quad \int_0^\omega k_x(u-v)\mathcal{L}_{r+1}(v)\,dv = u^r. \qquad (6.1.18)$$

Then the function

$$G_0(u) = \mathcal{F}_\tau(u) + \sum_{r=0}^n \gamma_r \mathcal{L}_{r+1}(u) \qquad (6.1.19)$$

satisfies (6.1.12).

Thus, the process of solving (6.1.12) reduces to a search for the functions $\mathcal{F}_\tau(u)$ and $\mathcal{L}_{r+1}(u)$ $(0 \le r \le n)$. From Chapter 2 it follows that all these functions can be expressed explicitly in terms of $\mathcal{L}_1(u)$ and $\mathcal{L}_2(u)$, if $k_x(u+\tau) \in C^{(2)}$ and

$$R = \int_0^\omega \mathcal{L}_1(u)\,du \ne 0. \qquad (6.1.20)$$

We note that the recurrence formulae from Chapter 1 can be used in order to determine $\mathcal{L}_r(u)$ $(r > 2)$. From (6.1.11) and (6.1.19) we obtain relations which determine the values γ_r:

$$\sum_{r=0}^n \gamma_r \int_0^\omega \mathcal{L}_{r+1}(u)u^p\,du = (-1)^p \tau^p - \int_0^\omega \mathcal{F}_\tau(u)u^p\,du, \quad 0 \le p \le n. \qquad (6.1.21)$$

For simplicity, we considered only the prediction problem. The filtration problem and other problems of linear optimal synthesis can be treated along the same lines [49].

Remark 6.1.2. Let the operator

$$Sf = \int_0^\omega k_x(u-v)f(v)\,dv$$

be such that $\langle Sf, f \rangle > 0$ for $f \ne 0$ $(f \in L(0, \omega))$. Then, the inequality

$$\det \left\{ \int_0^\omega \mathcal{L}_{r+1}(u)u^p\,du \right\}_{r,p=0}^n \ne 0$$

is valid. Hence, system (6.1.21) has one and only one solution $\gamma_0, \gamma_1, \ldots, \gamma_n$.

6.2   Diffraction on a strip

1. We consider a strip defined by the relations

$$0 \le x \le \omega, \quad -\infty < y < \infty, \quad z = 0. \qquad (6.2.1)$$

Suppose that a plane harmonic wave

$$\phi = e^{ik(x\sin\theta + z\cos\theta)}, \quad k = \overline{k}, \qquad (6.2.2)$$

falls into this strip, where θ is the angle between the direction of propagation of the wave and the normal to the strip (see Figure 6.1).

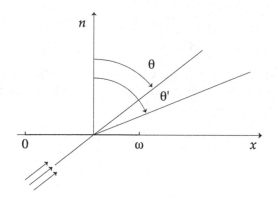

Figure 6.1: Difraction on a strip.

Then in the case of Dirichlet conditions, the scattered wave $v(x, z, \theta)$ can be found [24] from the equality

$$v(x, z, \theta) = -\frac{i}{4} \int_0^\omega \sigma(t, \theta) H_0^{(1)} \left(k\sqrt{(x - t)^2 + z^2} \right) dt, \qquad (6.2.3)$$

where $H_0^{(1)}$ is the Hankel function of the first kind, and σ is given by

$$\sigma(x, \theta) = \left. \frac{\partial v(x, z, \theta)}{\partial z} \right|_{z=-0}^{z=+0}$$

and satisfies the integral equation

$$\frac{i}{4} \int_0^\omega \sigma(t, \theta) H_0^{(1)} (k|x - t|) \, dt = e^{ikx \sin \theta}. \qquad (6.2.4)$$

Thus, the considered diffraction problem reduces to solving (6.2.4). The solvability of (6.2.4) is shown (as usual) by expanding $\sigma(t, \theta)$ in a series of Matthieu functions [25]. Let us prove this, starting from the general results obtained in Chapter 2.

Theorem 6.2.1. *If $\varphi'(x) \in L^p(0, \omega)$ $(1 < p \leq 2)$, then*

$$Sf = \frac{i}{4} \int_0^\omega f(t) H_0^{(1)} (k|x - t|) \, dt = \varphi(x) \qquad (6.2.5)$$

has one and only one solution in $L^r(0, \omega)$ $(1 < r < p)$.

Proof. We use the asymptotic behavior of the Hankel function as $x \to 0$:

$$H_0^{(1)} (k|x|) = -\frac{1}{2\pi} \ln |x| + h(x), \qquad (6.2.6)$$

where $h(x)$ satisfies for some c the inequality

$$|h'(x)| \leq c(\ln|x| + 1), \quad -\omega \leq x \leq \omega. \tag{6.2.7}$$

If $f \in L^p(0, \omega)$ $(1 < p \leq 2)$, then it follows from (6.2.5) and (6.2.6) that $Sf \in L^q(0, \omega)$. Consequently, the scalar product $\langle Sf, f \rangle$ is well defined for $f \in L^p(0, \omega)$. Taking into account the representation

$$H_0^{(1)}(k|x|) = \frac{1}{\pi} \int_{-\infty}^{\infty} \frac{1}{\sqrt{1 - y^2}} e^{ixyk} \, dy,$$

we obtain

$$\operatorname{Im}\langle Sf, f \rangle = \frac{1}{4\pi} \int_{-1}^{1} \frac{1}{\sqrt{1 - y^2}} \left| \int_{0}^{\omega} f(t) e^{-ikty} \, dt \right|^2 dy > 0 \text{ for } \|f\|_p \neq 0. \tag{6.2.8}$$

It follows from (6.2.8) that the equation

$$Sf = 0 \tag{6.2.9}$$

has only the trivial solution in $L^r(0, \omega)$ $(1 < r < 2)$. According to (6.2.6) and (6.2.7), the operator S satisfies the regularization conditions (2.5.3). Using Theorem 2.5.2 we complete the proof. □

2. According to Theorem 6.2.1, equation (6.2.4) has in $L^p(0, \omega)$ $(1 < p < 2)$ one and only one solution $\sigma(x, \theta)$. Furthermore, from Theorem 6.2.1 and Theorem 2.2.4 it follows that $\gamma = 0$. Using the notations from Chapter 2 we write

$$\begin{cases} \mathcal{L}_1(x) = \sigma(x, 0), & R = \int_0^\omega \sigma(x, 0) \, dx, \\ B(x, \lambda) = \sigma(x, \theta), & \lambda = k \sin \theta. \end{cases} \tag{6.2.10}$$

An important characteristic feature of the scattered field is the scattering diagram defined (see [24]) by

$$\varphi(\theta, \theta') = \int_0^\omega \sigma(t, \theta) e^{-ikt \sin \theta'} \, dt, \tag{6.2.11}$$

where θ' is the angle between the direction of the point of observation and the normal to the strip. In view of (6.2.4) and (6.2.11), we have

$$\varphi(\theta, \theta') = \langle \sigma(x, \theta), S\sigma(x, \theta') \rangle. \tag{6.2.12}$$

Hence,

$$\operatorname{Im}\varphi(\theta, \theta) = -\operatorname{Im}\langle S\sigma(x, \theta), \sigma(x, \theta) \rangle. \tag{6.2.13}$$

Comparing (6.2.8) and (6.2.13) we have

$$\operatorname{Im}\varphi(\theta,\theta) < 0. \tag{6.2.14}$$

From (6.2.10) and (6.2.14) we deduce

$$\operatorname{Im} R = \operatorname{Im}\int_0^\omega \sigma(x,0)\,\mathrm{d}x < 0, \quad\text{that is, }\ R \neq 0. \tag{6.2.15}$$

By virtue of Theorem 6.2.1, there is a function $\mathcal{L}_2(x)$ in $L^p(0,\omega)$ $(1 < p < 2)$ such that

$$S\mathcal{L}_2 = x. \tag{6.2.16}$$

It is easy to see that

$$\mathcal{L}_2(x) = \frac{1}{ik}\left.\frac{\partial\sigma(x,\theta)}{\partial\theta}\right|_{\theta=0}. \tag{6.2.17}$$

Taking into account (1.3.4)–(1.3.6) and (2.2.25), we see that the function $B(x,\lambda)$ can be expressed in terms of $\mathcal{L}_1(x)$ and $\mathcal{L}_2(x)$. Hence, $\sigma(x,\theta)$ can be expressed in terms of $\sigma(x,0)$ and $\left.\dfrac{\partial\sigma(x,\theta)}{\partial\theta}\right|_{\theta=0}$ by the following formulae:

$$\sigma(x,\theta) = u(x,\theta) - ik\sin\theta\int_x^\omega e^{ik\sin\theta(x-t)}u(t,\theta)\,\mathrm{d}t,$$

$$u(x,\theta) = \frac{a(\theta)}{R}\left(\int_x^\omega \sigma(t,0)\,\mathrm{d}t + \frac{1}{ik}\left.\frac{\partial\sigma(x,\theta)}{\partial\theta}\right|_{\theta=0}\right) + b(\theta)\sigma(x,0),$$

$$a(\theta) = ik\sin\theta\int_0^\omega e^{ik\sin\theta t}\sigma(\omega - t,0)\,\mathrm{d}t,$$

$$b(\theta) = \frac{1}{R}\int_0^\omega e^{ik\sin\theta t}\left(\sigma(\omega - t,0) + \sin\theta\left.\frac{\partial\sigma(\omega - t,\theta)}{\partial\theta}\right|_{\theta=0}\right)\mathrm{d}t.$$

Comparing (6.2.11) with (2.1.17) we obtain

$$\varphi(\theta,\theta') = \rho(\lambda,-\mu), \quad \lambda = k\sin\theta, \quad \mu = k\sin\theta'.$$

Now using (2.1.18) we have

$$\varphi(\theta,\theta') = -\frac{i}{k}\frac{a(\theta)b(\theta') - b(\theta)a(\theta')}{\sin\theta - \sin\theta'}e^{-ik\omega\sin\theta'}. \tag{6.2.18}$$

Formula (6.2.18) reveals the structure of the scattering diagram. It turns out that the function $\varphi(\theta,\theta')$ depending on two variables θ and θ' is completely determined by the two functions $a(\theta)$ and $b(\theta)$ of the single variable θ. The results of point 2 of Section 6.2 were obtained by Y.N. Lebedev and I.M. Polishchuk [58].

6.3   Extremal problems in the theory of synthesis of antennae

1. Let P be the operator acting from $L^2(0,\omega)$ to $L^2(-\infty,\infty)$ determined by

$$Pf = \int_0^\omega f(x)e^{i\lambda x}\,dx, \quad -\infty < \lambda < \infty. \tag{6.3.1}$$

The problem of synthesis of antennae in aperture approximation theory reduces to solving the equation

$$Pf + g = 0, \tag{6.3.2}$$

where $g(\lambda)$ is a given function in $L^2(-\infty,\infty)$. The equation (6.3.2) is solvable only under strong conditions on $g(\lambda)$. In particular, $g(\lambda)$ must be entire. Therefore, it is convinient to replace the problem of solving (6.3.2) exactly by that of minimizing in some sense the expression $Pf + g$.

Let us introduce a non-negative bounded weight function $\mu(\lambda)$ such that

$$\int_{-\infty}^\infty \mu(\lambda)\,d\lambda < \infty. \tag{6.3.3}$$

Next, we consider two related problems P1 and P2 [43, 84].

P1 Let some number $\varepsilon > 0$ and function $g(\lambda) \in L^2(-\infty,\infty)$ be given. Then, in the class of functions $f(x)$ such that

$$\|f\|^2 = \int_0^\omega |f(x)|^2\,dx \leq \varepsilon, \tag{6.3.4}$$

we should find a function $f_0(x)$ for which the functional

$$N(f) = \int_{-\infty}^\infty \mu(\lambda)|Pf + g|^2\,d\lambda \tag{6.3.5}$$

attains its minimal value.

P2 Let some number $\varepsilon > 0$ and function $g(\lambda) \in L^2(-\infty,\infty)$ be given. Then, in the class of functions $f(x) \in L^2(0,\omega)$ such that

$$N(f) \leq \varepsilon, \tag{6.3.6}$$

we should find a function $f_0(x)$ for which $\|f\|$ attains its minimal value.

We start with a simpler problem P3 related to P1 and P2.

P3 Suppose that a number $\nu > 0$ and a function $g(\lambda) \in L^2(-\infty, \infty)$ are given. It is required to find $f_\nu \in L^2(0, \omega)$ for which

$$E(f) = N(f) + \nu \|f\|^2 \tag{6.3.7}$$

attains its minimal value.

Problem P3 can be explained in the following way: we should find a good approximation f_ν for a solution of (6.3.2) and to diminish the norm of that approximation simultaneously. The weight function $\mu(\lambda)$ characterizes different requirements to the smallness of $Pf + g$ on different parts of the real axis.

Theorem 6.3.1. *There is one and only one function $f_\nu(x)$ which minimizes the functional $E(f)$. It is the solution of the integral equation*

$$\nu f_\nu(x) + \int_0^\omega f_\nu(t)k(x - t)\,dt = g_0(x), \quad 0 < x < \omega, \tag{6.3.8}$$

where

$$\begin{cases} k(x) = \int_{-\infty}^\infty \mu(\lambda)e^{-i\lambda x}\,d\lambda, \\ g_0(x) = -\int_{-\infty}^\infty \mu(\lambda)g(\lambda)e^{-i\lambda x}\,d\lambda. \end{cases} \tag{6.3.9}$$

Proof. The operator

$$Kf = \int_0^\omega f(t)k(x - t)\,dt$$

is non-negative, by (6.3.9). Hence, $\nu E + K$ is strictly positive. Thus, (6.3.8) has a unique solution $f_\nu(x)$.

We write an arbitrary function $f(x)$ in the form

$$f(x) = f_\nu(x) + h(x), \quad \text{where } h(x) = f(x) - f_\nu(x).$$

Then we find by a direct calculation that

$$E(f) = E(f_\nu) + \int_{-\infty}^\infty \mu(\lambda)|Ph|^2 d\lambda + \nu \|h\|^2, \tag{6.3.10}$$

and the theorem's statement follows. □

We note that we look for a solution of the problem P3 in the class of real-valued functions. We do not require $g(\lambda)$ to be real.

Theorem 6.3.2. *In the class of real-valued functions, there is one and only one function $f_\nu(x)$ for which the functional $E(f)$ attains its minimal value. This $f_\nu(x)$*

is the solution of the equation

$$\nu f_\nu(x) + \int\limits_0^\omega f_\nu(t) k(x-t)\, dt = g_0(x), \quad 0 \le x \le \omega, \tag{6.3.11}$$

where

$$\begin{cases} k(x) = \displaystyle\int_{-\infty}^\infty \mu(\lambda) \cos x\lambda\, d\lambda, \\[2mm] g_0(x) = -\displaystyle\int_{-\infty}^\infty \mu(\lambda) \mathrm{Re}\left(g(\lambda) e^{-i\lambda x}\right) d\lambda. \end{cases} \tag{6.3.12}$$

Proof. We write an arbitrary real-valued function $f(x)$ in the form

$$f(x) = f_\nu(x) + h(x), \tag{6.3.13}$$

where $f_\nu(x)$ is defined by (6.3.11) and (6.3.12), and $h(x) = f(x) - f_\nu(x)$. It is easy to check that (6.3.10) also holds in this case, and from this the theorem follows. $\square$

2. Now we proceed to solve problems P1 and P2. Our arguments for the class of real-valued functions and in the general case are the same.

We introduce the functions

$$\varphi(\nu) = \|f_\nu\|^2, \quad \varphi_N(\nu) = N(f_\nu), \tag{6.3.14}$$

where f_ν is defined by (6.3.8) and (6.3.9) in the general case and by (6.3.11) and (6.3.12) in the real-valued case.

Theorem 6.3.3. *The function $\varepsilon = \varphi(\nu)$ is strictly monotonically decreasing and the function $\varepsilon = \varphi_N(\nu)$ is strictly monotonically increasing on the semi-axis $0 < \nu < \infty$.*

Proof. Since the operator K is non-negative the following operator functions are strictly monotonically decreasing:

$$T_\nu = \left\langle (\nu E + K)^{-1}, T_\nu^2 \right\rangle.$$

Hence, $\varphi(\nu) = \langle T_\nu^2 g_0, g_0 \rangle$ is strictly monotonically decreasing.

Next we turn to $\varphi_N(\nu)$. Since

$$f_\nu = (\nu E + K)^{-1} g_0, \tag{6.3.15}$$

we deduce from (6.3.5) that

$$\varphi_N(\nu) = \int\limits_{-\infty}^\infty \mu(\lambda) |g(\lambda)|^2\, d\lambda - \left\langle (K + 2\nu E) f_\nu, f_\nu \right\rangle, \tag{6.3.16}$$

and now it follows immediately that

$$\varphi'_N(\nu) = 2\nu\Big\langle (K + \nu E)^{-1} f_\nu, f_\nu \Big\rangle > 0.$$

Thus, $\varphi_N(\nu)$ is strictly monotonically increasing. This proves the theorem. □

It follows from (6.3.15) and (6.3.16) that

$$\lim_{\nu \to +\infty} \varphi(\nu) = 0, \quad \lim_{\nu \to +\infty} \varphi_N(\nu) = \int_{-\infty}^{\infty} \mu(\lambda) |g(\lambda)|^2 \, d\lambda = I_g. \qquad (6.3.17)$$

Since $\varphi(\nu)$ and $\varphi_N(\nu)$ are monotone, the limits

$$\lim_{\nu \to +0} \varphi(\nu) = m_g, \quad \lim_{\nu \to +0} \varphi_N(\nu) = i_g,$$

exist, and

$$0 < m_g \leq \infty, \quad 0 \leq i_g < I_g. \qquad (6.3.18)$$

We denote by $\nu(\varepsilon)$ $(0 < \varepsilon < m_g)$, and $\nu_N(\varepsilon)$ $(i_g < \varepsilon < I_g)$ the inverse functions of $\varepsilon = \varphi(\nu)$ and $\varepsilon = \varphi_N(\nu)$.

Theorem 6.3.4. *If $0 < \varepsilon < m_g$, then the problem* P1 *has one and only one solution, namely,* $f_{\nu(\varepsilon)}(x)$.

Proof. We write the equation

$$\left\| f_{\nu(\varepsilon)}(x) \right\|^2 = \varphi(\nu(\varepsilon)) = \varepsilon. \qquad (6.3.19)$$

Thus, $f_{\nu(\varepsilon)}(x)$ satisfies (6.3.4). Now we take an arbitrary function $f(x)$ satisfying (6.3.4), but with $f(x) \neq f_{\nu(\varepsilon)}(x)$.

We put $\|f\|^2 = \beta$; then $\beta \leq \varepsilon$, by (6.3.4). Applying Theorem 6.3.1, or, in the case of real-valued $f(x)$, Theorem 6.3.2, we obtain

$$N(f) \geq N(f_{\nu(\beta)}). \qquad (6.3.20)$$

Since $\nu(\varepsilon)$ decreases monotonically,

$$\nu(\beta) \geq \nu(\varepsilon), \qquad (6.3.21)$$

and one of the inequalities (6.3.20) or (6.3.21) must be strict. Hence, we have the inequality $N(f) > N(f_{\nu(\varepsilon)})$. This proves the theorem. □

Similarly one proves the following result.

Theorem 6.3.5. *If $i_g < \varepsilon < I_g$, then the problem* P2 *has the unique solution* $f_{\nu_N(\varepsilon)}(x)$.

Remark 6.3.6. *If $\varepsilon > I_g$, then the problem* P2 *has the trivial solution $f(x) = 0$.*

Chapter 7

Lévy Processes: Convolution-type Form of the Infinitesimal Generator

7.1 Introduction

1. Let us introduce the notion of Lévy processes.

Definition 7.1.1. A stochastic process $\{X_t : t \geq 0\}$ is called a Lévy process, if the following conditions are fulfilled:

1. Almost surely $X_0 = 0$, that is, $P(X_0 = 0) = 1$.
   (One says that an event happens almost surely (a.s.) if it happens with probability one.)

2. For any $0 \leq t_1 < t_2 < \cdots < t_n < \infty$ the random variables

$$X_{t_2} - X_{t_1}, X_{t_3} - X_{t_2}, \ldots, X_{t_n} - X_{t_{n-1}} \tag{7.1.1}$$

   are independent (independent increments), that is, the increments (7.1.1) are mutually (not just pairwise) independent.

3. For any $s < t$ the distributions of $X_t - X_s$ and X_{t-s} are equal (stationary increments).

4. Process X_t is almost surely right continuous with left limits.

Then we have

$$\mu(z,t) = E\left[e^{izX_t}\right] = e^{-t\lambda(z)}, \quad t \geq 0, \tag{7.1.2}$$

where λ is given by the Lévy–Khinchine formula (see [8], [99])

$$\lambda(z) = \frac{1}{2}Az^2 - i\gamma z - \int\limits_{-\infty}^{\infty} \left(e^{ixz} - \mathbb{1} - ixz\mathbb{1}_{|x|<1}\right)\nu(\,\mathrm{d}x). \tag{7.1.3}$$

Here

$$A \geq 0, \quad \gamma = \overline{\gamma}, \quad z = \overline{z}, \quad \mathbb{1}_{|x|<1} = \begin{cases} 1, & \text{for} \quad |x| < 1 \\ 0, & \text{for} \quad |x| \geq 1 \end{cases}$$

and $\nu(\,\mathrm{d}x)$ is a measure on the axis $(-\infty, \infty)$ satisfying the conditions

$$\int\limits_{-\infty}^{\infty} \frac{x^2}{1+x^2} \nu(\,\mathrm{d}x) < \infty. \tag{7.1.4}$$

The Lévy process X_t is generated by the Lévy–Khinchine triplet $(A, \gamma, \nu(\,\mathrm{d}x))$. By $P_t(x_0, \Delta)$ we denote the probability $P(X_t \in \Delta)$ when $P(X_0 = x_0) = 1$ and $\Delta \subset \mathbb{R}$. The transition operator P_t is defined by the formula

$$P_t f(x) = \int\limits_{-\infty}^{\infty} P_t(x,\,\mathrm{d}y) f(y). \tag{7.1.5}$$

Let C_0 be the Banach space of continuous functions $f(x)$, satisfying the condition $\lim\limits_{|x| \to \infty} f(x) = 0$, with the norm $\|f\| = \sup_x |f(x)|$.
We denote by C_0^n the set of $f(x) \in C_0$ such that $f^{(k)}(x) \in C_0$ $(1 \leq k \leq n)$.
It is known [99] that if $f(x) \in C_0^2$ then

$$P_t f \in C_0. \tag{7.1.6}$$

Now, we formulate the following important result [99].

Theorem 7.1.2 (Lévy–Itō decomposition). *The family of operators P_t $(t \geq 0)$ defined by the Lévy process X_t is a strongly continuous semigroup on C_0 with norm $\|P_t\| = 1$. Let L be its infinitesimal generator. Then*

$$Lf = \frac{1}{2} A \frac{\mathrm{d}^2 f}{\mathrm{d}x^2} + \gamma \frac{\mathrm{d}f}{\mathrm{d}x} + \int\limits_{-\infty}^{\infty} \left(f(x+y) - f(x) - y \frac{\mathrm{d}f}{\mathrm{d}x} \mathbb{1}_{|y|<1} \right) \nu(\,\mathrm{d}y), \tag{7.1.7}$$

where $f \in C_0^2$.

Slightly changing the usual classification [99] we introduce the following definition:

Definition 7.1.3. We say that a Lévy process X_t generated by the triplet (A, ν, γ) belongs to class I if

$$A = 0 \quad \text{and} \quad \int\limits_{-\infty}^{\infty} \nu(\,\mathrm{d}x) < \infty, \tag{7.1.8}$$

and X_t belongs to class II if

$$A \neq 0 \quad \text{or} \quad \int_{-\infty}^{\infty} \nu(\mathrm{d}x) = \infty. \tag{7.1.9}$$

Remark 7.1.4. The introduced class I coincides with the class A in the usual classification. The introduced class II coincides with the union of the classes B and C in the usual classification.

The properties of these two classes of Lévy processes are quite different.

2. In the present chapter we show that the Lévy–Itō representation of the generator L can be written in the convolution form

$$Lf = \frac{\mathrm{d}}{\mathrm{d}x} S \frac{\mathrm{d}}{\mathrm{d}x} f, \tag{7.1.10}$$

where the operator S is defined by the relation

$$Sf = \frac{1}{2} Af + \int_{-\infty}^{\infty} k(y - x) f(y) \, \mathrm{d}y. \tag{7.1.11}$$

We note that for arbitrary a $(0 < a < \infty)$ the inequality

$$\int_{-a}^{a} |k(t)| \, \mathrm{d}t < \infty \tag{7.1.12}$$

is valid.

Formulas (7.1.10) and (7.1.11) were proved before in our works [97] under some additional conditions. In the present chapter we omit these additional conditions and prove these formulas for the general case. The representation of L in form (7.1.10) is convenient as the operator L is expressed via classic differential and convolution operators. This expression allows us to investigate the long time behavior of $p(t, \Delta)$ (see Chapters 8–10), where the notation $p(t, \Delta)$ stands for the probability that a sample of the process X_τ remains inside the domain Δ for $0 \leq \tau \leq t$ (ruin problem).

7.2 Convolution-type form of the infinitesimal generator

1. By $C(a)$ we denote the set of functions $f(x) \in C_0$ which have the following property:

$$f(x) = 0, \quad x \notin [-a, a], \tag{7.2.1}$$

that is, the function $f(x)$ is equal to zero in the neighborhood of $x = \infty$.

We introduce the functions

$$\mu_-(x) = \int_{-\infty}^{x} \nu(dx), \quad x < 0, \tag{7.2.2}$$

$$\mu_+(x) = -\int_{x}^{\infty} \nu(dx), \quad x > 0, \tag{7.2.3}$$

where the functions $\mu_-(x)$ and $\mu_+(x)$ are monotonically increasing and right continuous on the half-axis $(-\infty, 0]$ and $[0, \infty)$ respectively. We note that

$$\mu_+(x) \to 0, \quad x \to +\infty; \qquad \mu_-(x) \to 0, \quad x \to -\infty, \tag{7.2.4}$$

$$\mu_-(x) \ge 0, \quad x < 0; \qquad \mu_+(x) \le 0, \quad x > 0. \tag{7.2.5}$$

In view of (7.1.4) the integrals on the right-hand sides of (7.2.2) and (7.2.3) are convergent.

Theorem 7.2.1. *The following relations*

$$\varepsilon^2 \mu_\pm(\pm\varepsilon) \to 0, \quad \varepsilon \to +0, \tag{7.2.6}$$

$$\int_{-a}^{0} x\mu_-(x)\,dx < \infty, \quad -\int_{0}^{a} x\mu_+(x)\,dx < \infty, \quad 0 < a < \infty \tag{7.2.7}$$

are valid.

Proof. According to (7.1.4) we have

$$0 \le \int_{-a}^{-\varepsilon} x^2\,d\mu_-(x) \le M, \tag{7.2.8}$$

where M does not depend on ε. Integrating by parts the integral of (7.2.8) we obtain:

$$\int_{-a}^{-\varepsilon} x^2\,d\mu_-(x) = \varepsilon^2\mu_-(-\varepsilon) - a^2\mu_-(-a) - 2\int_{-a}^{-\varepsilon} x\mu_-(x)\,dx \le M. \tag{7.2.9}$$

The function $-\int_{-a}^{-\varepsilon} x\mu_-(x)\,dx$ of ε is monotonic increasing. In view of (7.2.9) this function is bounded. Hence, we have

$$\lim_{\varepsilon \to +0} \int_{-a}^{-\varepsilon} x\mu_-(x)\,dx = \int_{-a}^{0} x\mu_-(x)\,dx. \tag{7.2.10}$$

It follows from (7.2.9) and (7.2.10) that

$$\lim_{\varepsilon \to +0} \varepsilon^2 \mu_-(-\varepsilon) = m, \quad m \geq 0. \tag{7.2.11}$$

Using (7.2.10) and (7.2.11) we have $m = 0$. Thus, relations (7.2.9) and (7.2.10) are proved for $\mu_-(x)$. In the same way relations (7.2.9) and (7.2.10) can be proved for $\mu_+(x)$. □

2. Let us introduce the functions

$$k_-(x) = \int_{-b}^{x} \mu_-(t)\, dt, \quad -\infty \leq x < 0, \quad b > 0, \tag{7.2.12}$$

$$k_+(x) = -\int_{x}^{b} \mu_+(t)\, dt, \quad 0 < x \leq +\infty. \tag{7.2.13}$$

In view of (7.2.5) the integrals on the right-hand sides of (7.2.11) and (7.2.12) are absolutely convergent. From (7.2.11) and (7.2.12) we obtain the following statements.

Theorem 7.2.2.

1. *The function $k_-(x)$ is continuous, monotonically increasing on $(-\infty, 0)$ and*

$$k_-(x) \geq 0, \quad -b \leq x < 0. \tag{7.2.14}$$

2. *The function $k_+(x)$ is continuous, monotonically decreasing on the $(0, +\infty)$ and*

$$k_+(x) \geq 0, \quad 0 < x \leq b. \tag{7.2.15}$$

Further we need the following result.

Theorem 7.2.3. *The relations*

$$\varepsilon k_-(-\varepsilon) \to 0, \quad \varepsilon \to +0; \qquad \varepsilon k_+(\varepsilon) \to 0, \quad \varepsilon \to +0; \tag{7.2.16}$$

$$\int_{-b}^{0} k_-(x)\, dx < \infty; \qquad \int_{0}^{b} k_+(x)\, dx < \infty \tag{7.2.17}$$

are valid.

Proof. According to (7.2.5) we have

$$0 \leq -\int_{-b}^{-\varepsilon} x\mu_-(x)\, dx \leq M, \tag{7.2.18}$$

where M does not depend on ε. Integrating the integral in (7.2.18) by parts, we obtain:

$$-\int_{-b}^{-\varepsilon} x\mu_-(x)\,\mathrm{d}x = \varepsilon k_-(-\varepsilon) - k_-(-b) + \int_{-b}^{-\varepsilon} k_-(x)\,\mathrm{d}x \le M. \qquad (7.2.19)$$

The function $\displaystyle\int_{-b}^{-\varepsilon} k_-(x)\,\mathrm{d}x$ is monotonic increasing with respect to ε. This function is bounded (see (7.2.19)). Hence, we have

$$\lim_{\varepsilon\to+0}\int_{-b}^{-\varepsilon} k_-(x)\,\mathrm{d}x = \int_{-b}^{0} k_-(x)\,\mathrm{d}x. \qquad (7.2.20)$$

It follows from (7.2.19) and (7.2.20) that

$$\lim_{\varepsilon\to+0}\varepsilon k_-(-\varepsilon) = p, \quad p \ge 0. \qquad (7.2.21)$$

Using (7.2.20) and (7.2.21) we have $p = 0$. Thus, relations (7.2.16) and (7.2.17) are proved for $k_-(x)$. In the same way relations (7.2.16) and (7.2.17) can be proved for $k_+(x)$. $\qquad\square$

3. We use the following notation

$$J(f) = J_1(f) + J_2(f), \qquad (7.2.22)$$

where

$$J_1(f) = \frac{\mathrm{d}}{\mathrm{d}x}\int_{-\infty}^{x} f'(y)k_-(y-x)\,\mathrm{d}y, \quad f(x) \in C(a), \qquad (7.2.23)$$

$$J_2(f) = \frac{\mathrm{d}}{\mathrm{d}x}\int_{x}^{\infty} f'(y)k_+(y-x)\,\mathrm{d}y, \quad f(x) \in C(a). \qquad (7.2.24)$$

Lemma 7.2.4. *The operator $J(f)$ defined by (7.2.22) can be represented in the form*

$$J(f) = \int_{-\infty}^{\infty}\left(f(y+x) - f(x) - y\frac{\mathrm{d}f(x)}{\mathrm{d}x}\mathbb{1}_{|y|\le1}\right)\mu(\mathrm{d}y) + \Gamma f'(x), \qquad (7.2.25)$$

where $\Gamma = \overline{\Gamma}$ and $f(x) \in C(a)$.

Proof. From (7.2.24) we obtain the relation

$$J_1(f) = -\int_{x-1}^{x} (f'(y) - f'(x))k'_-(y-x)\,dy - \int_{-a}^{x-1} f'(y)k'_-(y-x)\,dy. \quad (7.2.26)$$

Here we used relations (7.2.12) and the equality

$$\int_{x-1}^{x} k_-(y-x)\,dy = \int_{-1}^{0} k_-(v)\,dv. \quad (7.2.27)$$

We introduce the notations

$$P_1(x,y) = f(y) - f(x) - (y-x)f'(x), \quad P_2(x,y) = f(y) - f(x). \quad (7.2.28)$$

Using notations (7.2.28) we rewrite (7.2.26) in the form

$$J_1(f) = -\int_{-1}^{0} \frac{\partial}{\partial y} P_1(x, y+x)\mu_-(y)\,dy - \int_{-a-x}^{-1} \frac{\partial}{\partial y} P_2(x, y+x)\mu_-(y)\,dy. \quad (7.2.29)$$

Integrating by parts the integrals of (7.2.29) we deduce that

$$J_1(f) = f'(x)\gamma_1 + \int_{-1}^{0} P_1(x, y+x)\,d\mu_-(y)$$

$$+ \int_{-a-x}^{-1} P_2(x, y+x)\,d\mu_-(y) + P_2(x, -a)\mu_-(-a), \quad (7.2.30)$$

where $\gamma_1 = k'_-(-1)$. It follows from (7.1.4) that the integrals in (7.2.30) are absolutely convergent. Passing to the limit in (7.2.30), when $a \to +\infty$, and taking into account (7.2.16), (7.2.27) we have

$$J_1(f) = \int_{-\infty}^{x} \left(f(y+x) - f(x) - y\frac{df(x)}{dx}\mathbb{1}_{|y|\leq 1} \right) d\mu_-(y) + \gamma_1 f'(x). \quad (7.2.31)$$

In the same way it can be proved that

$$J_2(f) = \int_{x}^{\infty} \left(f(y+x) - f(x) - y\frac{df(x)}{dx}\mathbb{1}_{|y|\leq 1} \right) d\mu_+(y) + \gamma_2 f'(x), \quad (7.2.32)$$

where $\gamma_2 = k'_+(1)$. The relation (7.2.26) follows directly from (7.2.31) and (7.2.32). Here $\Gamma = \gamma_1 + \gamma_2$. The lemma is proved. $\qquad\square$

Remark 7.2.5. The operator $L_0 f = \dfrac{\mathrm{d}}{\mathrm{d}x} f$ can be represented in form (7.1.10), (7.1.11), where

$$S_0 f = \int\limits_{-\infty}^{\infty} p_0(x - y) f(y)\, \mathrm{d}y, \qquad (7.2.33)$$

$$p_0(x) = \frac{1}{2}\, \mathrm{sgn}\,(x). \qquad (7.2.34)$$

From Lemma 7.2.4, and Remark 7.2.5 we deduce the following assertion.

Theorem 7.2.6. *The infinitesimal generator L has a convolution-type form* (7.1.10), (7.1.11).

4. Now, we shall show that formula (7.1.10) can be simplified if the Lévy process X_t belongs to the class I. Without loss of generality we assume that

$$\gamma - \int\limits_{|y|<1} y\nu(\,\mathrm{d}y) = 0. \qquad (7.2.35)$$

Remark 7.2.7. If condition (7.2.35) is valid, then class I coincides with the class of compound Poisson processes.

According to (7.1.7) the generator L of the corresponding process X_t can be represented in the form

$$L f = \int\limits_{-\infty}^{\infty} \big(f(x + y) - f(x)\big)\nu(\,\mathrm{d}y). \qquad (7.2.36)$$

Using Theorem 7.2.6 we have:

Proposition 7.2.8. *Let the Lévy process X_t belong to the class I and let the condition* (7.2.35) *be fulfilled. Then formula* (7.1.7) *can be written in the following convolution form*

$$L f = \frac{\mathrm{d}}{\mathrm{d}x} S \frac{\mathrm{d}}{\mathrm{d}x} f, \qquad (7.2.37)$$

where

$$S f = \int\limits_{-\infty}^{\infty} k(y - x) f(y)\, \mathrm{d}y, \qquad (7.2.38)$$

$$k(x) = \begin{cases} k_+(x), & if \ \ x > 0, \\ k_-(x), & if \ \ x < 0. \end{cases} \qquad (7.2.39)$$

Using (7.2.15) and (7.2.16) we obtain the following assertion.

Proposition 7.2.9. *Let the Lévy process X_t belong to the class* I *and conditions* (7.2.35) *be fulfilled. Then relation* (7.1.10) *takes the form*

$$Lf = -\Omega f + \int_{-\infty}^{\infty} f(y) \, d_y \nu(y - x), \quad \Omega = \int_{-\infty}^{\infty} d\nu(y). \tag{7.2.40}$$

Proof. It follows from (7.2.15) and (7.2.16) that

$$Lf = -\int_{-\infty}^{x} \mu_-(y - x) f'(y) \, dy - \int_{x}^{\infty} \mu_+(y - x) f'(y) \, dy. \tag{7.2.41}$$

Integrating (7.2.41) by parts we have

$$Lf = -\big(\mu_-(0) - \mu_+(0)\big) f + \int_{-\infty}^{\infty} f(y) \, d_y \nu(y - x). \tag{7.2.42}$$

The proposition is proved. □

In formulas (7.2.40) and (7.2.42) we use the equality $d\nu(x) = \nu(\,dx)$.

Definition 7.2.10. We say that Lévy process X_t belongs to class I_c if the corresponding Lévy measure $\nu(y)$ is summable and continuous.
We say that Lévy process X_t belongs to class I_d if the corresponding Lévy measure $\nu(y)$ is summable and discrete.

It is easy to see that the following assertion holds.

Proposition 7.2.11. *If Lévy process X_t belongs to the class* I, *then X_t can be represented in the form*

$$X_t = X_t^{(1)} + X_t^{(2)}, \tag{7.2.43}$$

where $X_t^{(1)} \in I_c$ and $X_t^{(2)} \in I_d$.

Example 7.2.12. Let us consider the Poisson process. In this case we have

$$\lambda(z) = -\big(e^{iz} - 1\big). \tag{7.2.44}$$

According to (7.1.3) and (7.2.44) the relations

$$\mu_-(x) = 0, \quad x < 0; \qquad \mu_+(x) = \begin{cases} -1, & \text{if } 0 < x < 1, \\ 0, & \text{if } x \geq 1 \end{cases} \tag{7.2.45}$$

are valid. Using (7.2.8), (7.2.9) and (7.2.36) we obtain

$$k_-(x) = 0, \quad x < 0; \qquad k_+(x) = \begin{cases} 1 - x, & \text{if } 0 < x < 1, \\ 0, & \text{if } x \geq 1. \end{cases} \tag{7.2.46}$$

Hence, the operator L for the Poisson process has the following convolution form:

$$Lf = \frac{\mathrm{d}}{\mathrm{d}x} \int\limits_{x}^{x+1} (1 - y + x)f'(y)\,\mathrm{d}y. \tag{7.2.47}$$

Formula (7.2.47) coincides with the Lévy–Itō formula:

$$Lf = -f'(x) + f(x + 1) - f(x). \tag{7.2.48}$$

Chapter 8

On the Probability that the Lévy Process (Class II) Remains within the Given Domain

8.1 Introduction

Recall that by $p(t, \Delta)$ we denote the probability that a sample of the process X_τ remains inside the domain Δ for $0 \leq \tau \leq t$ (ruin problem). Here Δ is the set of the segments $[a_k, b_k]$, where

$$a_1 < b_1 < a_2 < b_2 < \cdots < a_n < b_n \quad (1 \leq k \leq n). \tag{8.1.1}$$

It is easy to see that, if T_Δ is the time during which X_τ remains in the domain Δ before it leaves the domain Δ for the first time, we have

$$p(t, \Delta) = P(T_\Delta > t). \tag{8.1.2}$$

In this chapter we consider in detail the Lévy processes of class II. We construct a quasi-potential operator B, which plays an essential role in our approach to the Lévy process theory. This operator B is linear and bounded in the space of continuous functions. With the help of the operator B we find a new formula for $p(t, \Delta)$:

$$\int_0^\infty e^{-st} p(t, \Delta) \, dt = \int_\Delta d_x \Psi_\infty(x, s), \quad s > 0. \tag{8.1.3}$$

The definition of $\Psi_\infty(x, s)$ is given in Section 8.3. Formula (8.1.3) allows us to obtain the long time behavior of $p(t, \Delta)$. Namely, we proved the following asymptotic formula:

$$p(t, \Delta) = e^{-t/\lambda_1}(c_1 + o(1)), \quad c_1 > 0, \quad \lambda_1 > 0, \quad t \to +\infty. \tag{8.1.4}$$

8.2 Support

Let us recall some basic definitions (see [99]).

Definition 8.2.1. The support $S(\rho)$ of the measure ρ on $\mathbb{R}$ is the set of points $x \in \mathbb{R}$ such that $\rho(G) > 0$ for any interval G containing x.

The support $S(\rho)$ is a closed set.

Definition 8.2.2. For any random variable X on $\mathbb{R}$ the support of the corresponding distribution $F(x)$ is called the support of X and is denoted by $S(X)$.

Now we can formulate the well-known H. Tucker theorem [107].

Theorem 8.2.3. *Let X_t be a Lévy process on $\mathbb{R}$ with a triplet (A, ν, γ).*

1) *If either $A > 0$ or*

$$\int_{-\infty}^{\infty} |x| \, d\nu = \infty, \tag{8.2.1}$$

then $S(X_t) = \mathbb{R}$.

2) *If $0 \in S(\nu)$, $S(\nu) \bigcap (0, \infty) \neq 0$, $S(\nu) \bigcap (-\infty, 0) \neq 0$, then $S(X_t) = \mathbb{R}$.*

3) *Suppose that $A = 0$, $0 \in S(\nu)$ and*

$$\int_{-\infty}^{\infty} |x| \, d\nu < \infty. \tag{8.2.2}$$

If $S(\nu) \in [0, \infty)$, then $S(X_t) = [t\gamma, \infty)$.
If $S(\nu) \in (-\infty, 0]$, then $S(X_t) = (-\infty, t\gamma]$.

Further, we suppose that the following conditions are fulfilled:

$$X_0 = 0, \quad 0 \in \Delta \subset S(\nu). \tag{8.2.3}$$

8.3 Lévy processes (class II)

1. In many theoretical and applied problems it is important to estimate the value $p(t, \Delta) = P(X_\tau \in \Delta; 0 \leq \tau \leq t)$, that is, the probability that a sample of the process X_τ remains inside Δ for $0 \leq \tau \leq t$ (*ruin problem*). In this section we investigate the properties of $p(t, \Delta)$.

Recall that we consider in this chapter only the Lévy processes from class II. We denote by $F_0(x, t)$ the distribution function of Lévy process X_t, that is,

$$F_0(x, t) = P(X_t \leq x). \tag{8.3.1}$$

We need the following statement [99, p. 175, Theorem 27.4].

Theorem 8.3.1. *The distribution function $F_0(x,t)$ is continuous with respect to x if and only if the Lévy process belongs to class* II.

We introduce the sequence of functions

$$F_{n+1}(x,t) = \int\limits_{0}^{t} \int\limits_{-\infty}^{\infty} F_0(x-\xi, t-\tau)V(\xi)\,\mathrm{d}_\xi F_n(\xi,\tau)\,\mathrm{d}\tau, \qquad (8.3.2)$$

where the function $V(x)$ is defined by relations $V(x) = \begin{cases} 1 & \text{for } x \notin \Delta \\ 0 & \text{for } x \in \Delta \end{cases}$. In the right-hand side of (8.3.2) we use Stieltjes integration. It follows from (7.1.2) that

$$\mu(z,t) = \mu(z,t-\tau)\mu(z,\tau). \qquad (8.3.3)$$

Due to (8.3.3) and a convolution formula for the Stieltjes–Fourier transform (see [9, Chap. 4]), the relation

$$F_0(x,t) = \int\limits_{-\infty}^{\infty} F_0(x-\xi, t-\tau)\,\mathrm{d}_\xi F_0(\xi,\tau) \qquad (8.3.4)$$

is true. Using (8.3.2) and (8.3.4) we have

$$0 \le \mathrm{d}_x F_n(x,t) \le \frac{t^n\,\mathrm{d}_x F_0(x,t)}{n!}, \quad \text{if } \mathrm{d}x > 0. \qquad (8.3.5)$$

Relation (8.3.5) implies that

$$0 \le F_n(x,t) \le \frac{t^n F_0(x,t)}{n!}. \qquad (8.3.6)$$

Hence, the series

$$F(x,t,u) = \sum_{n-0}^{\infty} (-1)^n u^n F_n(x,t) \qquad (8.3.7)$$

converges. The probabilistic meaning of $F(x,t,u)$ is shown by the relation (see [31, Chap. 4]):

$$E\left\{ \exp\left(-u\int\limits_{0}^{t} V(X_\tau)\,\mathrm{d}\tau\right), c_1 < X_t < c_2 \right\} = F(c_2,t,u) - F(c_1,t,u). \qquad (8.3.8)$$

The inequality $V(x) \ge 0$ and relation (8.3.8) imply that the function $F(x,t,u)$ monotonically decreases with respect to the variable u and monotonically increases with respect to the variable x. Hence, the formula

$$0 \le F(x,t,u) \le F(x,t,0) = F_0(x,t) \qquad (8.3.9)$$

is valid. In view of (8.3.1) and (8.3.9) the Laplace transform

$$\Psi(x, s, u) = \int_0^\infty e^{-st} F(x, t, u)\, dt, \quad s > 0 \tag{8.3.10}$$

is well defined. According to (8.3.2) the function $F(x, t, u)$ is the solution of the equation

$$F(x, t, u) + u \int_0^t \int_{-\infty}^\infty F_0(x - \xi, t - \tau) V(\xi)\, d\xi F(\xi, \tau, u)\, d\tau = F_0(x, t). \tag{8.3.11}$$

Applying a Laplace transform to both parts of (8.3.11), taking into account (8.3.10) and using the convolution property [9, Chap. 4], we obtain

$$\Psi(x, s, u) + u \int_{-\infty}^\infty \Psi_0(x - \xi, s) V(\xi)\, d\xi \Psi(\xi, s, u) = \Psi_0(x, s), \tag{8.3.12}$$

where

$$\Psi_0(x, s) = \int_0^\infty e^{-st} F_0(x, t)\, dt. \tag{8.3.13}$$

It follows from (7.1.2) and (8.3.13) that

$$\int_{-\infty}^\infty e^{ixp}\, d_x \Psi_0(x, s) = \frac{1}{s + \lambda(p)}. \tag{8.3.14}$$

According to (8.3.12) and (8.3.13) we have

$$\int_{-\infty}^\infty e^{ixp} \big(s + \lambda(p) + u V(x)\big)\, d_x \Psi(x, s, u) = 1. \tag{8.3.15}$$

By C_Δ we denote the set of twice-differentiable functions $g(x)$ on $\mathbb{R}$ such that

$$g(a_k) = g(b_k) = g'(a_k) = g'(b_k) = 0, \quad 1 \le k \le n \tag{8.3.16}$$

and $g''(x)$ is continuous. Now we introduce the function

$$h(p) = \frac{1}{2\pi} \int_\Delta e^{-ixp} f(x)\, dx, \tag{8.3.17}$$

where $f(x)$ belongs to C_Δ. Multiplying both parts of (8.3.16) by $h(p)$ and integrating them with respect to p ($-\infty < p < \infty$) we deduce the equality

$$\int_{-\infty}^{\infty} \int_{-\infty}^{\infty} e^{ixp}(s + \lambda(p)) h(p) \, d_x \Psi(x, s, u) \, dp = f(0). \tag{8.3.18}$$

We have used the relations

$$V(x)f(x) = 0, \quad -\infty < x < \infty, \tag{8.3.19}$$

$$\frac{1}{2\pi} \lim \int_{-N}^{N} \int_{\Delta} e^{-ixp} f(x) \, dx \, dp = f(0), \quad N \to \infty. \tag{8.3.20}$$

Since the function $F(x, t, u)$ monotonically decreases with respect to u, this is also true for the function $\Psi(x, s, u)$ (see (8.3.10)). Hence, there exist the limits

$$F_\infty(x, t) = \lim F(x, t, u), \quad \Psi_\infty(x, s) = \lim \Psi(x, s, u), \quad u \to \infty. \tag{8.3.21}$$

It follows from (8.3.8) (see [32, Section 8]) that

$$p(t, \Delta) = P\big(X_\tau \in \Delta, \, 0 < \tau < t\big) = \int_\Delta d_x F_\infty(x, t). \tag{8.3.22}$$

Hence, we have

$$\int_0^{\infty} e^{-st} p(t, \Delta) \, dt = \int_\Delta d_x \Psi_\infty(x, s), \, s > 0. \tag{8.3.23}$$

Using relations (7.1.3) and (7.1.7) we deduce that

$$\lambda(z) \int_{-\infty}^{\infty} e^{-iz\xi} f(\xi) \, d\xi = -\int_{-\infty}^{\infty} e^{-iz\xi} (Lf)(\xi) \, d\xi. \tag{8.3.24}$$

2. Introduce the operator P_Δ by the relations

$$P_\Delta f(x) = \begin{cases} f(x) & \text{for} \quad x \in \Delta, \\ 0 & \text{for} \quad x \notin \Delta. \end{cases}$$

We note that the equality $P_\Delta \dfrac{d}{dx} = \dfrac{d}{dx} P_\Delta$, for $x \in \Delta$, yields the relation

$$P_\Delta L P_\Delta = \frac{d}{dx} S_\Delta \frac{d}{dx}, \quad \text{where} \quad S_\Delta = P_\Delta S P_\Delta.$$

Definition 8.3.2. The operator

$$L_\Delta := P_\Delta L P_\Delta = \frac{d}{dx} S_\Delta \frac{d}{dx} \quad (S_\Delta := P_\Delta S P_\Delta) \tag{8.3.25}$$

is called a truncated generator.

Relations (8.3.20), (8.3.21) and (8.3.24) imply the following assertion.

Theorem 8.3.3. *Let X_t be a Lévy process. If the corresponding distribution function $F(x,t)$ is continuous with respect to x, then for $s > 0$ we have the equality*

$$\int_\Delta (sI - L_\Delta) f \, d_x \Psi_\infty(x,s) = f(0) \quad \text{for} \quad f \in C_\Delta. \tag{8.3.26}$$

Remark 8.3.4. For the symmetric stable processes, equality (8.3.26) was deduced by M. Kac (see [31, p. 204, formula (8.8)]). For the Lévy processes with continuous density, (8.3.26) was obtained in [97, p. 24, Theorem 1.31]. Here we deduced equality (8.3.26) for the case of Lévy processes with continuous distribution.

3. Let us consider the behavior of $\Psi_\infty(x,s)$ when $s = 0$. For that purpose we need the following Hengartner and Thedorescu result (see [23], [99, p. 363, Lemma 48.3]):

Theorem 8.3.5. *Let X_t be a Lévy process. Then for any finite interval K the asymptotics*

$$P(X_t \in K) = O(t^{-1/2}) \quad \text{as} \quad t \to \infty \tag{8.3.27}$$

is valid.

Hence, we have the assertion.

Theorem 8.3.6. *Let X_t be a Lévy process. Then for any integer $n > 0$ the asymptotics*

$$p(t, \Delta) = O(t^{-n/2}) \quad \text{as} \quad t \to \infty \tag{8.3.28}$$

is valid.

Proof. Let the interval K be such, that $\Delta \subset K$ and $\Delta - \Delta \subset K$. Then the inequality

$$P(X_t \in \Delta) \le P(X_t \in K) \tag{8.3.29}$$

holds. According to Lévy processes properties (independent and stationary increments) and (8.3.29), the inequality

$$p(t, \Delta) \le P\left(X_{t/n} \in \Delta\right) \prod_{j=2}^n P\left((X_{tj/n} - X_{t(j-1)/n}) \in \Delta - \Delta\right) \le P\left(X_{t/n} \in \Delta\right)^n \tag{8.3.30}$$

is valid. Relations (8.3.27) and (8.3.30) imply relation (8.3.28). □

It follows from (8.3.10) that

$$\int_0^\infty e^{-st} p(t, \delta)\, dt = \int_\delta d_x \Psi_\infty(x, s), \qquad (8.3.31)$$

where δ is a set of segments which belong to Δ and

$$p(t, \delta) = P(X_\tau \in \delta;\, 0 \le \tau \le t). \qquad (8.3.32)$$

If $\delta = \Delta$, then formula (8.3.31) coincides with formula (8.3.23). We need the following partial case of (8.3.31):

$$\int_0^\infty p(t, \delta)\, dt = \int_\delta d_x \Psi_\infty(x, 0), \qquad (8.3.33)$$

According to (8.3.28) the integral in the left-hand side of (8.3.33) exists. Let us prove the following statement.

Proposition 8.3.7. *The function $\Psi_\infty(x, 0)$ is monotonically increasing and continuous in the domain Δ.*

Proof. If $\delta \subset \Delta$, then $p(t, \delta) \le p(t, \Delta)$. It follows from this fact and from (8.3.33) that the function $\Psi_\infty(x, 0)$ is monotonically increasing in the domain Δ. Let us prove that the function $\Psi_\infty(x, 0)$ is continuous.

The function $F(x, t, u)$ is continuous with respect to x. Hence, in view of (8.3.10) the function $\Psi(x, s, u)$ is continuous with respect to x, when $s > 0$. Using relations (8.3.8) and (8.3.21) we see that the functions $F(x, t, u)$ and $\Psi(x, s, u)$ uniformly converge, when $u \to \infty$. Hence, the functions $\Phi_\infty(x, t)$ and $\Psi_\infty(x, s)$, $(s > 0)$ are continuous with respect to x. Passing to the limit in equality (8.3.31), when $s \to 0$, and using (8.3.28) we obtain that the function $\Psi_\infty(x, s)$ uniformly converges. Then the function $\Psi_\infty(x, 0)$ is continuous with respect to x. The proposition is proved. $\square$

(See also Point 9 in Commentaries and Remarks for additional explanations to Proposition 8.3.7.)

8.4 Quasi-potential

1. The operator

$$Qf = \int_0^\infty (P_t f)\, dt$$

is called the potential of the semigroup P_t. $\Longleftarrow$ +par

We note that the operator P_t is defined by relation (7.1.5) and

$$-LQf = f, \quad f \in C_0^2.$$

We recall that the domain Δ is the set of segments $[a_k, b_k]$, where

$$a_1 < b_1 < a_2 < b_2 < \cdots < a_n < b_n, \quad 1 \leq k \leq n.$$

We denote by D_Δ the space of the continuous functions $g(x)$ on the domain Δ. We denote by D_Δ^0 the subspace of the continuous functions $g(x)$ on the domain Δ such that

$$g(a_k) = g(b_k) = 0, \quad 1 \leq k \leq n. \tag{8.4.1}$$

The norms in D_Δ and in D_Δ^0 are defined by the relation $\|g\| = \sup_{x \in \Delta} |g(x)|$.

Definition 8.4.1. The operator B with the domain of definition D_Δ is called a *quasi-potential* if the following relation is valid:

$$-BL_\Delta g = g, \quad g \in C_\Delta. \tag{8.4.2}$$

Remark 8.4.2. In a number of cases (see the next section) we need relation (8.4.2). In these cases we can use the quasi-potential B, which is often simpler than the corresponding potential Q.

Theorem 8.4.3. *Let the considered Lévy process X_t belong to the class II. Then there exists a quasi-potential B of the form*

$$Bf = \int_\Delta f(y) \, d_y \Phi(x, y), \quad f \in D_\Delta, \tag{8.4.3}$$

where the real-valued function $\Phi(x, y)$ is continuous with respect to x and y, and monotonically increasing with respect to y.

Before proving Theorem 8.4.3, we study some properties of the operator B, which is introduced by (8.4.3). The operator B maps the space D_Δ into itself. We recall the following definitions.

Definition 8.4.4. The total variation of a complex-valued function g, defined on segment $[a_k, b_k]$, is the value

$$V_k(g) = \sup_P \sum_{i=0}^{n_P - 1} |g(x_{i+1}) - g(x_i)|,$$

where the supremum is taken over the set of all partitions $P = (x_0, x_1, \ldots, x_{n_P})$ of the segment $[a_k, b_k]$.

Definition 8.4.5. The total variation of a complex-valued function g, defined on the domain Δ is the value

$$V_\Delta = \sum_{k=1}^n V_k. \tag{8.4.4}$$

Definition 8.4.6. A complex-valued function g on the Δ is said to be of bounded variation on the Δ if its total variation is finite.

By D_Δ^* we denote the space, which is adjoint to D_Δ. It is well known that the space D_Δ^* consists of functions $g(x)$ with the bounded total variation $V_\Delta(g)$. The norm in D_Δ^* is defined by the relation $\|g\| = V_\Delta(g)$, the functional in D_Δ is defined by the relation

$$\langle f, g \rangle_\Delta = \int_\Delta f(x)\, d\overline{g(x)}, \quad f \in D_\Delta, \quad g \in D_\Delta^*. \tag{8.4.5}$$

Hence, the adjoint operator B^* maps the space D_Δ^* into D_Δ^* and has the form

$$B^* g = \int_\Delta \Phi(y, x)\, dg(y). \tag{8.4.6}$$

The weight function

$$\sigma(y) \equiv \begin{cases} -1/2, & \text{when } y < 0, \\ 1/2, & \text{when } y > 0 \end{cases}$$

has only one point of growth at $y = 0$, and Stieltjes integration in (8.4.6) gives us the equality

$$B^* \sigma = \Phi(0, x)\big(\sigma(+0) - \sigma(-0)\big) = \Phi(0, x). \tag{8.4.7}$$

The next assertion follows directly from (8.3.26).

Lemma 8.4.7. *If $y_0 = 0$ belongs to the inner part of Δ and $\Phi(0, x) = \Psi_\infty(x, 0)$, then*

$$-\langle L_{\Delta(0)} f, B^* \sigma \rangle_\Delta = f(0). \tag{8.4.8}$$

Condition 8.4.8. *Further we assume that 0 is an inner point of Δ. When we consider y_0 we assume that y_0 also is an inner point of Δ.*

Let us consider an arbitrary inner point $y_0 \in \Delta$. We introduce the new domain $\Delta(y_0) = \Delta - y_0$. The corresponding truncated generator is denoted by $L_{\Delta(y_0)}$, the corresponding quasi-potential is denoted by $B(y_0)$, the corresponding kernel is denoted by $\Phi(x, y, y_0)$, and the corresponding Ψ-function is denoted by $\Psi_\infty(x, y, y_0)$.

Now, we shall reduce the general case to the case $y_0 = 0$. We introduce the operator

$$Uf = g(x), \quad g(x) = f(x - y_0), \quad x \in \Delta, \tag{8.4.9}$$

which maps the space $D_{\Delta(y_0)}$ onto D_Δ. Using formula (8.4.9) we deduce that

$$L_\Delta = U L_{\Delta(y_0)} U^{-1}. \tag{8.4.10}$$

Hence, the equality

$$B = U B(y_0) U^{-1} \tag{8.4.11}$$

is valid. The last equality can be rewritten in the terms of the kernels:

$$\Phi(x,y) = \Phi(x - y_0, y - y_0, y_0). \tag{8.4.12}$$

Relation (8.3.26) in the case $\Delta(y_0)$ takes the form

$$\int\limits_{\Delta(y_0)} -L_{\Delta(y_0)} f \, d_x \Psi_\infty(x, 0, y_0) = f(0). \tag{8.4.13}$$

According to (8.4.13) Lemma 8.4.7 can be written in the following form:

Lemma 8.4.9. *If* $\Phi(0, x, y_0) = \Psi_\infty(x, 0, y_0)$, *then*

$$-\big\langle L_{\Delta(y_0)} f, B(y_0)^* \sigma \big\rangle_{\Delta(y_0)} = f(0). \tag{8.4.14}$$

In view of (8.4.10) and (8.4.11) equality (8.4.14) can be rewritten in the form

$$\big\langle -L_{\Delta(0)} g, B(0)^* \sigma(x - y_0) \big\rangle_\Delta = g(y_0), \tag{8.4.15}$$

where

$$g(x) = f(x + y_0). \tag{8.4.16}$$

Using (8.4.14) we define the kernel $\Phi(x, y)$ of the operator B by the relation

$$\Phi(x, y) = \Psi_\infty(y - x, 0, x). \tag{8.4.17}$$

According to Proposition 8.3.7 and (8.4.17) we have the following assertion.

Proposition 8.4.10. *The function* $\Phi(x, y)$ *is continuous with respect to* x *and* y *and monotonically increasing with respect to* y. *The corresponding operator* B *is bounded in the space* D_Δ.

Relations (8.4.17) and (8.3.26) imply the equality

$$-BL_\Delta g = g, \quad g \in C_\Delta. \tag{8.4.18}$$

Now, we can prove Theorem 8.4.3.

Proof of Theorem 8.4.3. It follows from (8.4.18) that the constructed operator B is quasi-potential. The theorem is proved. □

2. Sectorial properties. We shall need the following Pringsheim result.

Theorem 8.4.11 (see [104, Ch.1]). *Let the function* $f(t)$ *be non-increasing over* $(0, \infty)$ *and integrable on any finite interval* $(0, \ell)$. *If* $f(t) \to 0$ *for* $t \to \infty$, *then for any positive* x *we have*

$$\frac{1}{2}\big(f(x+0) + f(x-0)\big) = \frac{2}{\pi} \int\limits_{+0}^{\infty} \cos(xu) \left(\int\limits_0^\infty f(t)\cos(tu)\,dt \right) du, \tag{8.4.19}$$

$$\frac{1}{2}\big(f(x+0) + f(x-0)\big) = \frac{2}{\pi} \int\limits_0^{\infty} \sin(xu) \left(\int\limits_0^\infty f(t)\sin(tu)\,dt \right) du. \tag{8.4.20}$$

Recall that $\mu_{\pm}$ are given by formulas (7.2.12) and (7.2.13). We set

$$k_-(x) = \int_{-b}^{x} \mu_-(t)\, dt, \quad -b \leq x < 0, \tag{8.4.21}$$

$$k_+(x) = -\int_{x}^{b} \mu_+(t)\, dt, \quad 0 < x \leq b. \tag{8.4.22}$$

Hence, we have

$$k_-(-x) = k_+(x) = 0, \quad x \geq b; \tag{8.4.23}$$

$$k(x) = \begin{cases} k_+(x) & \text{for} \quad x > 0, \\ k_-(x) & \text{for} \quad x < 0. \end{cases} \tag{8.4.24}$$

The integral part of the operator S we denote by K, that is,

$$S_\Delta f = \frac{1}{2} A f + K_\Delta f, \quad K_\Delta := P_\Delta K P_\Delta, \quad K := \int_{-\infty}^{\infty} k(y-x) \cdot dy. \tag{8.4.25}$$

Using integration by parts we deduce the assertion below.

Proposition 8.4.12. *Let conditions (8.4.21) and (8.4.22) be fulfilled. Then we have*

$$\int_{-b}^{b} k(t) \cos(xt)\, dt = -\left(k'_+(b) - k'_-(-b)\right) \frac{1 - \cos(bx)}{x^2} + \int_{-b}^{b} \frac{1 - \cos(xt)}{x^2} \, d\mu(t),$$

$$\tag{8.4.26}$$

where $\mu(t) = \mu_-(t)$ for $t < 0$ and $\mu(t) = \mu_+(t)$ for $t > 0$.

The relations

$$k'_-(-b) = \mu_-(-b) \geq 0, \quad k'_+(b) = \mu_+(b) \leq 0 \tag{8.4.27}$$

are valid. It follows from (8.4.26) and Theorem 8.4.11 that the kernel $k(x)$ admits the representation

$$k(x) = \frac{1}{2\pi} \int_{-\infty}^{\infty} m(t) e^{ixt}\, dt, \tag{8.4.28}$$

where

$$\text{Re}\,(m(x)) = -\left(k'_+(b) - k'_-(-b)\right) \frac{1 - \cos(bx)}{x^2} + \int_{-b}^{b} \frac{1 - \cos(xt)}{x^2}\, d\mu(t) \geq 0. \tag{8.4.29}$$

Further we need the following notions.

Definition 8.4.13. The linear operator T with dense domain $D(T)$ in the space $L^2(\Delta)$ is called sectorial if

$$\langle Tf, f \rangle \neq 0, \quad f \in D(T), \quad f \neq 0 \tag{8.4.30}$$

and

$$-\frac{\pi}{2}\beta \leq \arg\langle Tf, f \rangle \leq \frac{\pi}{2}\beta, \quad 0 < \beta \leq 1. \tag{8.4.31}$$

Definition 8.4.14. The sectorial operator T is called strongly sectorial if the relation (8.4.31) is valid for some $\beta < 1$.

We note that we use two different scalar product-type notations, namely,

$$\langle f, g \rangle = \int\limits_\Delta f(x)\overline{g(x)}\, \mathrm{d}x \quad \text{and} \quad \langle f, g \rangle_\Delta = \int\limits_\Delta f(x)\, \mathrm{d}\overline{g(x)}.$$

Due to (8.4.24) and (8.4.25) the relation

$$\langle K_\Delta f, f \rangle = \int\limits_{-\infty}^{\infty} m(u) \left| \int\limits_\Delta f(t) e^{iut}\, \mathrm{d}t \right|^2 \mathrm{d}u \tag{8.4.32}$$

is valid. Since $0 \in \Delta$, the Lévy measure is positive (i.e., $\nu(\Delta) > 0$). In this case the entire function $m_1(x) = \operatorname{Re} m(x)$ is equal to zero only in finite number of points on every finite interval. In view of (8.4.32) we have $\langle K_\Delta f, f \rangle \neq 0$ and

$$\operatorname{Re}\langle K_\Delta f, f \rangle > 0 \quad \text{for} \quad f \neq 0. \tag{8.4.33}$$

Thus, in view of (8.4.33) we derived the following assertion.

Proposition 8.4.15. *If the Lévy process X_t belongs to the class* II *and* 0 *is the inner point of Δ, then the corresponding operator S_Δ is sectorial, that is,*

$$-\frac{\pi}{2} < \arg\langle S_\Delta f, f \rangle < \frac{\pi}{2}, \quad f \in L^2(\Delta),\ f \neq 0. \tag{8.4.34}$$

3. Strongly sectorial

Proposition 8.4.16. *If the operator S_Δ is self-adjoint then this operator is strongly sectorial.*

If $A = 0$, then $S_\Delta = K_\Delta$. Using relation (8.4.32) we obtain the assertion.

Proposition 8.4.17. *If for some $N > 0$ the inequality*

$$\operatorname{Re} m(u) \geq N|\operatorname{Im} m(u)|, \quad -\infty < u < \infty \tag{8.4.35}$$

is valid, then the corresponding operator S_Δ is strongly sectorial.

Proposition 8.4.18. *If*

$$k(x) = |x|^{1-\alpha}(1 - \beta(\operatorname{sgn} x)), \quad 0 < \alpha < 2, \quad \alpha \neq 1, \quad -1 \le \beta \le 1, \quad (8.4.36)$$

then the corresponding operator S_Δ is strongly sectorial.

Proof. In case (8.4.36) we have

$$m(u) = \left(\cos\frac{\pi\alpha}{2} - i\beta\left(\sin\frac{\pi\alpha}{2}\right)\operatorname{sgn}(u)\right)\frac{|u|^{\alpha-2}}{|\Gamma(\alpha - 1)\sin(\pi\alpha)|}. \quad (8.4.37)$$

We see that the condition (8.4.35) is fulfilled. The proposition is proved. □

Proposition 8.4.19. *Let the following conditions be fulfilled:*

1. *For some $m > 0$ the inequality*

$$\frac{m}{|x|^2} \le \nu'(x), \quad |x| \le 1 \quad (8.4.38)$$

   holds.
2. *For some $M > 0$ the inequality*

$$\left|\int_{-b}^{b} k(t)\sin(xt)\,dt\right| \le \frac{M}{|x|}, \quad |x| \le 1 \quad (8.4.39)$$

   holds.

Then the corresponding operator S_Δ is strongly sectorial.

Proof. From formulas (8.4.26) and (8.4.38), (8.4.39) we conclude that

$$\int_{-b}^{b} k(t)\cos(xt)\,dt \ge \int_{-1/|x|}^{1/|x|} \nu'(x)\frac{1-\cos(xt)}{x^2}\,dt \ge \frac{N}{|x|}, \quad N > 0, \quad |x| \ge 1. \quad (8.4.40)$$

It follows from (8.4.39) and (8.4.40) that the corresponding operator S_Δ is strongly sectorial. □

Proposition 8.4.20. *If the number λ belongs to the spectrum of the sectorial operator T then $\operatorname{Re}\lambda \ge 0$.*

Proposition 8.4.21. *If the number λ belongs to the spectrum of the strongly sectorial operator T then λ belongs to the domain (8.4.31), where $0 < \beta < 1$.*

We note that the operators S_Δ are essential not only for Lévy processes but also for other theories (see [92, 97]).

Example 8.4.22 (The variance damped Lévy process). In this case we have

$$\nu'(x) = C_1 e^{-\eta_1|x|}|x|^{-\alpha-1}\mathbb{1}_{x<0} + C_2 e^{-\eta_2 x}x^{-\alpha-1}\mathbb{1}_{x>0}, \qquad (8.4.41)$$

where $C_1 \geq 0$, $C_2 \geq 0$, $C_1 + C_2 \geq 0 > 0$, $\lambda_1 \geq 0$, $\lambda_2 \geq 0$, $0 < \alpha < 2$.

If $C_1 = C_2$ and $\alpha \geq 1$ then the conditions of Proposition 8.4.18 are fulfilled and the corresponding operator is strongly sectorial.

4. Stable Lévy processes. The Lévy measure in case of a stable Lévy process has the form (8.4.41) with $\eta_1 = \eta_2 = 0$. Using formulas (7.2.2), (7.2.3) and (7.2.12), (7.2.13) we obtain:

$$k(x) = a|x|^{1-\alpha}(1 - \beta\,\mathrm{sgn}\,x), \quad \text{where} \quad a = \frac{C_1 + C_2}{2\alpha(\alpha-1)}, \quad \beta = \frac{C_1 - C_2}{C_1 + C_2}. \quad (8.4.42)$$

It is easy to see that $-1 \leq \beta \leq 1$. In view of Proposition 8.4.17 we obtain the following assertion.

Corollary 8.4.23. *If the Lévy process is stable with parameter $0 < \alpha < 2$, $\alpha \neq 1$, then the corresponding operator S_Δ is strongly sectorial.*

Remark 8.4.24. In the case $0 < \alpha < 1$ we suppose that $\displaystyle\int_\Delta g(t)\,\mathrm{d}t = 0$.

If the Lévy process is stable with parameters $\alpha = 1$, $\beta = 0$, then

$$k(x) = -\ln\frac{|x|}{b}, \quad |x| \leq b. \qquad (8.4.43)$$

According to Proposition 8.4.18 we have

Corollary 8.4.25. *If the Lévy process is stable with parameters $\alpha = 1$, $\beta = 0$, then the corresponding operator S_Δ is strongly sectorial.*

5. Sectorial properties, the operator L_Δ. Using (7.1.10) we have

$$\langle L_\Delta g, g \rangle = -\langle S_\Delta g', g' \rangle, \qquad (8.4.44)$$

where the function g has the second continuous derivative and $g \in D_\Delta^0$. Relation (8.4.44) implies the following statement.

Proposition 8.4.26.

1. *If the operator S_Δ is sectorial, then the corresponding operator $(-L_\Delta)$ is sectorial too.*

2. *If the operator S_Δ is strongly sectorial, then the corresponding operator $(-L_\Delta)$ is strongly sectorial too.*

6. Dual Lévy process $\widetilde{X}_t$, and equality $\widetilde{L}_\Delta = L_\Delta^*$. We introduce the following notion.

Definition 8.4.27. Let X_t be a Lévy process with Lévy measure $\nu(dx)$. The Lévy process $\widetilde{X}_t$ with Lévy measure $\widetilde{\nu}(dx)$ is called dual to X_t if

$$A = \widetilde{A}, \quad \gamma = -\widetilde{\gamma}, \quad \mu(x) = -\widetilde{\mu}(-x), \tag{8.4.45}$$

where we use relations (7.2.2), (7.2.3) and equalities $\mu(x) = \mu_+(x)$, $x > 0$ and $\mu(x) = \mu_-(x)$, $x < 0$.

The notion of the dual process $\widetilde{X}_t$ can be given in terms of distribution function of process X_t (see [99, p. 286]):

Proposition 8.4.28. *Let $\widetilde{X}_t$ be a dual Lévy process to Lévy process X_t. Then the corresponding operators $\widetilde{L}_\Delta$, L_Δ and $\widetilde{S}_\Delta$, S_Δ are connected by the relations*

$$\widetilde{L}_\Delta = L_\Delta^*, \quad \widetilde{S}_\Delta = S_\Delta^*. \tag{8.4.46}$$

Proof. Formulas (7.2.12), (7.2.13) and (8.4.45) imply that $k(x) = \widetilde{k}(-x)$. Hence, we have $\widetilde{S}_\Delta = S_\Delta^*$. Taking into account equality (8.3.25) and conditions

$$g(a_k) = g(b_k) = 0 \quad (1 \leq k \leq n), \tag{8.4.47}$$

we have (8.4.46). $\qquad\square$

Corollary 8.4.29. *The operators L_Δ and $\widetilde{L}_\Delta$ generate a strongly continuous semigroups P_t $(t > 0)$ and $\widetilde{P}_t$ $(t > 0)$ respectively. The operators P_t $(t > 0)$ and $\widetilde{P}_t$ $(t > 0)$ are contraction operators in the Hilbert space $L^2(\Delta)$.*

Proof. The operators $-L_\Delta$ and $-L_\Delta^*$ are sectorial and densely defined. Hence, the assertion of the corollary is valid (see [54, p. 687]). $\qquad\square$

We note that the semigroup P_t is defined by relation (7.1.5).

Corollary 8.4.30. *The range of the operator L_Δ is dense in the Hilbert space $L^2(\Delta)$.*

Proof. The operator L_Δ^* is sectorial. Hence, the relation $L_\Delta^* f \neq 0$ is valid, where $f \neq 0$ and f belongs to the domain of definition of L_Δ^*. This proves the assertion of the corollary. $\qquad\square$

7. Sectorial properties, quasi-potential B. It follows from (8.4.36) that

$$-L_\Delta Bg = g, \quad g = L_\Delta f, \quad f \in C_\Delta. \tag{8.4.48}$$

Using (8.4.44), (8.4.48) and Corollary 8.4.30, we deduce the statement.

Proposition 8.4.31.

1. *If the operator S_Δ is sectorial, then the corresponding operator B is sectorial too.*

2. *If the operator S_Δ is strongly sectorial, then the corresponding operator B is strongly sectorial too.*

8. Semi-inner product, dissipative operators, strong maximum principle. A semi-inner product $[x, y]$ is defined in D_Δ and satisfies the following properties [54]:

$$[x + y, z] = [x, z] + [y, z], \quad [\lambda x, y] = \lambda [x, y], \tag{8.4.49}$$

$$[x, x] > 0 \quad \text{for} \quad x \neq 0, \quad \left| [x, y] \right|^2 \leq [x, x] [y, y], \tag{8.4.50}$$

where x, y, z belong to C_0^2 and λ is a complex number. Let us consider a linear, not necessarily bounded, operator T in the space C_0^2 with dense domain $\mathrm{Dom}(T)$.

Definition 8.4.32. An operator T is called dissipative if

$$\mathrm{Re}\,[Tx, x] \leq 0, \quad x \in \mathrm{Dom}(T). \tag{8.4.51}$$

We shall prove that the generating operator L and the operator L_Δ, defined by formulas (7.1.7) and (8.3.25), are dissipative. To do it we use the following **Maximum principle** [108, Chap. 4].

Proposition 8.4.33. *If the real-valued continuous function f has an absolute maximum in the point x (i.e., $|f(x)| \geq |f(y)|$ for all y) then*

$$Lf(x) \leq 0. \tag{8.4.52}$$

It follows from (8.3.25) that

$$L_\Delta f(x) \leq 0, \quad f \in D_\Delta. \tag{8.4.53}$$

Remark 8.4.34. The maximum principle for the Lévy processes follows directly from representation (7.1.7). We take into account that, in the maximal point x, we have $f'(x) = 0$, $f''(x) \leq 0$.

Now, let us consider complex-valued continuous function f which has an absolute maximum in the point x (i.e., $|f(x)| \geq |f(y)|$ for all y). We define the semi-inner product in D_Δ by relation

$$[g, f] = g(x)\overline{f(x)}. \tag{8.4.54}$$

It is obvious that relations (8.4.49) and (8.4.50) are fulfilled.

Proposition 8.4.35. *The operator L_Δ is dissipative.*

Proof. Without loss of generality we suppose that function f in the maximum point x is positive. Using (8.4.54) we obtain

$$\mathrm{Re}\,[L_\Delta f, f] \leq 0. \tag{8.4.55}$$

$\square$

Theorem 8.4.36. *If T is dissipative and $\operatorname{Re}\lambda > 0$, then $(\lambda I - T)^{-1}$ exists and*

$$\left\|(\lambda I - T)^{-1}\right\| \leq \frac{1}{\operatorname{Re}\lambda}. \tag{8.4.56}$$

Proof. We put $f = \lambda y - Ty$ and obtain

$$\operatorname{Re}\lambda[y,y] \leq \operatorname{Re}\lambda[y,y] - [Ty,y] = \operatorname{Re}[f,y] \leq \|f\|\,\|y\|,$$

that is,

$$\|y\| = \left\|(\lambda I - T)^{-1}f\right\| \leq \frac{\|f\|}{\operatorname{Re}\lambda}.$$

The theorem is proved. $\qquad\square$

For the case $\lambda > 0$, Theorem 8.4.36 was proved in the paper [54].

9. M. Kac's question. M. Kac (see [30, Formula (8.8)], [92]) obtained formula (8.3.26) only for the stable Lévy processes. He formulated the following problem and hypothesis:

Does relation (8.3.26) determine $\Psi_\infty(x, s)$ uniquely? The answer is undoubtedly "no", but we have no proof.

We show that in the general case (and not just for the stable processes) the answer is "yes".

Proposition 8.4.37. *Let the considered Lévy process X_t belong to class II and let 0 be the inner point of Δ. Then relation (8.3.26) determines $\Psi_\infty(x, s)$ uniquely.*

Proof. Indeed, the operators L_Δ and L_Δ^* are dissipative in the space $L^2(\Delta)$. Hence, the range of the operator $sI - L_\Delta$ ($s > 0$) coincides with $L^2(\Delta)$ (see [54, p. 687]). Using the relation $D_\Delta \subset L^2(\Delta)$ we prove the proposition. $\qquad\square$

Remark 8.4.38. The functions $g_1(x)$ and $g_2(x)$ are equivalent in the space D_Δ if $g_1(x) - g_2(x) = \text{const}$ for all $x \in \Delta$.

Taking into account the relation $D_\Delta \subset L^2(\Delta)$ and the Lumer–Phillips Theorem [54, Theorem 3.1] we obtain the following assertion.

Corollary 8.4.39. *Let conditions of Proposition 8.4.37 be fulfilled. Then the corresponding operator L_Δ generates a strongly continuous semigroup of contraction operators.*

8.5 Long time behavior

1. In this section we investigate the asymptotic behavior of $p(t, \Delta)$ when $t \to \infty$. Recall that according to Theorem 8.4.3 the operator B has the form (8.4.3) and is bounded in the space D_Δ.

Theorem 8.5.1. *Let the following conditions be fulfilled.*

   I. *The Lévy process X_t belongs to the class* II *and* 0 *is the inner point of Δ.*

  II. *The corresponding operator B is compact in the space D_Δ.*

Then we have:

   1. *The spectrum of the corresponding operator B belongs to the domain* Re $\lambda \geq 0$.

   2. *The function*

$$\Psi(x, s) = (I + sB^*)^{-1}\Phi(0, x) \tag{8.5.1}$$

   belongs to the space D_Δ^* and satisfies the relation (8.3.26).

Proof. The operator B is sectorial in the space $L^2(\Delta)$ (see Proposition 8.4.31). The eigenvalues and eigenfunctions of the compact operator B in the space D_Δ are eigenvalues and eigenfunctions of the sectorial operator B in the space $L^2(\Delta)$. According to Proposition 8.4.20 this spectrum belongs to the domain Re $\lambda \geq 0$. Hence, assertion 1 of the theorem is proved.

In view of (8.4.2) we have

$$-BL_\Delta f = f, \quad f \in C_\Delta. \tag{8.5.2}$$

Relations (8.5.1) and (8.5.2) imply that

$$\langle (sI - L_\Delta)f, \psi(x, s) \rangle_\Delta = -\langle (I + sB)L_\Delta f, \psi \rangle_\Delta = -\langle L_\Delta f, \Phi(0, x) \rangle_\Delta. \tag{8.5.3}$$

Since $\Phi(0, x) = B^*\sigma(x)$ (see (8.4.7)), according to (8.5.1) and (8.5.3) relation (8.3.26) is valid. $\qquad\square$

It follows from relations (8.3.26), (8.5.3) and Proposition 8.4.37 that

$$\Psi_\infty(x, s) = (I + sB^*)^{-1}\Phi(0, x). \tag{8.5.4}$$

Remark 8.5.2. We emphasize an important fact: operators B and B^* are bounded in the spaces D_Δ and D_Δ^*, respectively.

We apply the following Kreĭn–Rutman theorem [45, Section 6]:

Theorem 8.5.3. *If a linear compact operator B that maps a cone K into itself has a point of the spectrum different from zero, then it has a positive eigenvalue λ_1 not less in modulus than any other eigenvalue λ_k ($k > 1$). To this eigenvalue λ_1 corresponds at least one eigenvector $g_1 \in K$ of the operator B ($Bg_1 = \lambda_1 g_1$) and at least one eigenvector $h_1 \in K^*$ of the operator B^* ($B^* h_1 = \lambda_1 h_1$).*

We note that in our case the cone K consists of non-negative continuous real functions $g(x) \in D_\Delta$ and the cone K^* consists of monotonically increasing bounded functions $h(x) \in D_\Delta^*$. In view of Theorem 8.5.1 the non-zero eigenvalues $\{\lambda_k\}$ $(k \geq 1)$ of the operator B belong to the sector

$$-\frac{\pi}{2} \leq +\arg(z) \leq \frac{\pi}{2}, \quad |z| \leq \lambda_1. \tag{8.5.5}$$

Further, we assume that the following condition is fulfilled:

The non-zero eigenvalues $\{\lambda_k\}$ $(k \geq 1)$ of the operator B are situated in the sector

$$-\frac{\pi}{2}\beta \leq \arg(z) \leq \frac{\pi}{2}\beta, \quad 0 \leq \beta < 1, \quad |z| \leq \lambda_1. \tag{8.5.6}$$

Using the Kreĭn–Rutman theorem we have:

Theorem 8.5.4. *Let the following conditions be fulfilled.*

1. *Lévy process X_t belongs to the class* II.
2. *The corresponding quasi-potential B is compact in the Banach space D_Δ and has a point of the spectrum different from zero.*

Then the operator B has a positive eigenvalue λ_1 not less in modulus than any other eigenvalue λ_k $(k > 1)$. To this eigenvalue λ_1 corresponds at least one eigenfunction $g_1(x)$, which is continuous and non-negative, and at least one eigenfunction $h_1(x)$ of the operator B^*, which is monotonic and bounded.

We recall the following notions:

The *index* of the eigenvalue (ind λ_k) is defined as the dimension of the largest Jordan block associated with this eigenvalue.

The *geometric multiplicity* of the eigenvalue (gmul λ_k) is a number of linearly independent eigenfunctions corresponding to this eigenvalue.

Further, we assume that

$$\text{ind}\,\lambda_1 = \text{gmul}\,\lambda_1 = 1. \tag{8.5.7}$$

It follows from (8.5.7) that $\langle g_1, h_1 \rangle_\Delta > 0$ (see [45, Section 6]). We introduce the normalizing condition

$$\langle g_1, h_1 \rangle_\Delta = \int_\Delta g_1(x)\,dh_1(x) = 1. \tag{8.5.8}$$

Now, we formulate the main result of this section.

Theorem 8.5.5. *Let the following conditions be fulfilled.*

1. *The conditions of the Theorem 8.5.4 are fulfilled.*
2. *The relations (8.2.3) and (8.5.8) are valid.*

3. *The non-zero eigenvalues $\{\lambda_k\}$ $(k \geq 1)$ of the operator B are situated in the sector (8.5.6).*

4. *For any $m > 0$ there exists such M that*

$$\left|\langle (B - zI)^{-1}\mathbb{1}, \Phi(0, x)\rangle_\Delta\right| \leq \frac{M}{|z|^{m+1}}, \quad \frac{(\beta + \varepsilon)\pi}{2} \leq |\arg z| \leq \frac{(1 - \varepsilon)\pi}{2}, \quad \varepsilon > 0. \tag{8.5.9}$$

Then the asymptotic equality

$$p(t, \Delta) = e^{-t/\lambda_1}\left(c_1 + o(1)\right), \quad t \to +\infty, \quad c \geq 0. \tag{8.5.10}$$

holds. (If $g_1(0) > 0$ then $c_1 > 0$.)

Proof. Using (8.3.23) we obtain the equality

$$p(t, \Delta) = \frac{1}{2\pi} \int\limits_{-\infty}^{\infty} \langle e^{iyt}, \Psi_\infty(x, iy)\rangle_\Delta \, dy, \quad t > 0. \tag{8.5.11}$$

Changing the variable $z = i/y$ and taking into account (8.5.4) we rewrite (8.5.11) in the form

$$p(t, \Delta) = \frac{1}{2i\pi} \int\limits_{-i\infty}^{i\infty} \langle e^{-t/z}, (zI - B^*)^{-1}\Phi(0, x)\rangle_\Delta \frac{dz}{z}, \quad t > 0. \tag{8.5.12}$$

We introduce the domain D_ε:

$$-\frac{\pi}{2}(\beta + \varepsilon) \leq \arg(z) \leq \frac{\pi}{2}(\beta + \varepsilon), \quad \left|z - \frac{1}{2}\lambda_1\right| < \frac{1}{2}(\lambda_1 - r), \tag{8.5.13}$$

where $0 < \varepsilon < 1 - \beta$, $\quad 0 < r < \lambda_1$. If z belongs to the domain D_ε then the relation

$$\operatorname{Re}\frac{1}{z} > \frac{1}{\lambda_1} \tag{8.5.14}$$

holds. Indeed, relation (8.5.13) is equivalent to the inequality

$$\left(x - \frac{\lambda_1}{2}\right)^2 + y^2 < \frac{\lambda_1^2}{4} \quad (x = \operatorname{Re} z, \; y = \operatorname{Im} z), \quad \text{that is,} \quad \left|z - \frac{\lambda_1}{2}\right| < \frac{\lambda_1}{2}.$$

We take r so small that the circle

$$\left|z - \frac{1}{2}\lambda_1\right| = \frac{1}{2}(\lambda_1 - r) \tag{8.5.15}$$

has the points z_1 and $z_2 = \overline{z_1}$ of the intersections with the half-lines

$$\arg(z) = \frac{\pi}{2}(\beta + \varepsilon) \quad \text{and} \quad \arg(z) = -\frac{\pi}{2}(\beta + \varepsilon).$$

We denote the boundary of domain D_ε by Γ_ε. We emphasize that Γ_ε contains only that part of circle (8.5.15) that is situated between the points z_1 and z_2. Without loss of generality we may assume that the eigenvalues $\lambda_k \neq 0$ do not belong to Γ_ε. Since B is compact, only a finite number of eigenvalues λ_k $(1 < k \leq m)$ of this operator does not belong to the domain D_ε. We deduce from formula (8.5.11) the relation

$$p(t, \Delta) = \sum_{k=1}^{m} \sum_{j=0}^{n_k-1} e^{-t/\lambda_k} t^j c_{k,j} + J, \tag{8.5.16}$$

where n_k is the index of the eigenvalue λ_k and

$$J = -\frac{1}{2i\pi} \int_{\Gamma_\varepsilon} \frac{1}{z} e^{-t/z} \langle \mathbb{1}, (B^* - zI)^{-1} \Phi(0, x) \rangle_\Delta \, dz. \tag{8.5.17}$$

Taking into account (8.5.7) we have

$$n_1 = 1. \tag{8.5.18}$$

Relation (8.4.7) implies that

$$\Phi(0, x) \in D_\Delta^*. \tag{8.5.19}$$

Among the numbers λ_k we choose the ones for which $\mathrm{Re}\,(1/\lambda_k)$ $(1 \leq k \leq m)$ has the smallest value δ. Among the obtained numbers we choose μ_k $(1 \leq k \leq \ell)$ the indexes n_k of which have the largest value n. We deduce from (8.5.9), (8.5.16) and (8.5.17) that

$$p(t, \Delta) = e^{-t\delta} t^n \big(Q(t) + o(1) \big) \quad (t \to \infty), \quad Q(t) := \sum_{k=1}^{\ell} e^{it\mathrm{Im}\,(\mu_k^{-1})} c_k. \tag{8.5.20}$$

We note that the function $Q(t)$ is almost periodic [50]. Hence, in view of (8.5.20) and the inequality $p(t, \Delta) \geq 0$ the relation

$$Q(t) \geq 0, \quad -\infty < t < \infty \tag{8.5.21}$$

is valid. First, we assume that the inequality

$$\delta < \lambda_1^{-1} \tag{8.5.22}$$

holds. Using (8.5.22) and the inequality

$$\lambda_1 \geq |\lambda_k|, \quad k = 2, 3, \ldots \tag{8.5.23}$$

we have

$$\mathrm{Im}\,(\mu_j^{-1}) \neq 0, \quad 1 \leq j \leq \ell. \tag{8.5.24}$$

It follows from the second equality in (8.5.20) that

$$c_j = \lim \frac{1}{2T} \int_{-T}^{T} Q(t)e^{-it\mathrm{Im}\,(\mu_j^{-1})}\,dt, \quad T \to \infty. \tag{8.5.25}$$

In view of (8.5.21) and (8.5.25) the relations

$$|c_j| \le \lim \frac{1}{2T} \int_{-T}^{T} Q(t)\,dt = 0, \quad T \to \infty \tag{8.5.26}$$

are valid, that is, $c_j = 0$ ($1 \le j \le \ell$). Hence, the equalities

$$\delta = \lambda_1^{-1}, \quad n = 1 \tag{8.5.27}$$

hold. From (8.5.27) we obtain the asymptotic equality

$$p(t, \Delta) = e^{-t/\lambda_1}(c_1 + o(1)), \quad t \to \infty, \tag{8.5.28}$$

where

$$c_1 = g_1(0) \int_{\Delta} dh_1(x) \ge 0. \tag{8.5.29}$$

The theorem is proved. $\qquad\square$

According to Theorem 8.5.5 and to the relation

$$0 < \mathrm{Re}\,\frac{1}{\lambda_k} \le \frac{1}{\lambda_1}$$

the following assertion holds.

Corollary 8.5.6. *Let the conditions of Theorem 8.5.5 be fulfilled. Then all the eigenvalues λ_j of B belong to the disk*

$$\left| z - \frac{1}{2}\lambda_1 \right| \le \frac{1}{2}\lambda_1. \tag{8.5.30}$$

2. Now, we assume that the function $\Phi(x, y)$ is absolutely continuous with respect to y. Hence, the corresponding operator B has the form

$$Bf = \int_{\Delta} f(y)K(x, y)\,dy, \quad K(x, y) = \frac{\partial}{\partial y}\Phi(x, y), \tag{8.5.31}$$

where for all $x \in \Delta$ the inequality

$$\int_{\Delta} |K(x, y)|\,dy < \infty \tag{8.5.32}$$

holds. In addition, we assume that for all $x \in \Delta$ the following relation

$$\lim_{h \to 0} \int_\Delta |K(x+h,y) - K(x,y)| \, dy = 0 \qquad (8.5.33)$$

is fulfilled. Then the operator B is compact in the space D_Δ. Here we use the well-known Radon theorem [61]:

Theorem 8.5.7. *The operator B of the form* (8.5.31) *is compact in the space D_Δ if and only if the relations* (8.5.32) *and* (8.5.33) *are fulfilled.*

3. We can formulate the condition (8.5.9) of Theorem 8.5.5 in terms of the kernel $K(x,y)$. We suppose that there exists such positive, monotonically decreasing function $r(s)$ that

$$|K(x,y)| \le r(|x-y|), \quad \int_0^b r^2(s)\, ds < \infty, \quad b > 0. \qquad (8.5.34)$$

It follows from (8.5.33), that the operator B is bounded in the Hilbert space $L^2(\Delta)$.

Let us introduce the following notion:

Definition 8.5.8. The numerical range $W(B)$ of the operator B in the Hilbert space $L^2(\Delta)$ is the set $W(B) = \{\langle Bf, f \rangle, \|f\| = 1\}$.

If the closure of the convex hull of $W(B)$ is situated in the sector (8.5.6), then the corresponding operator B is strongly sectorial in the Hilbert space $L^2(\Delta)$.

Proposition 8.5.9. *Let the conditions* (8.5.33) *be fulfilled. If the corresponding operator B is strongly sectorial in the Hilbert space $L^2(\Delta)$, then the estimate*

$$\|(B - zI)^{-1}\mathbb{1}\| \le \frac{M}{|z|}, \quad z \in \Gamma_\varepsilon \qquad (8.5.35)$$

is valid (see [103]).

It follows from (8.5.34) and (8.5.35) that

$$\left|\langle (B - zI)^{-1}\mathbb{1}, K(0,x) \rangle\right| \le \frac{M}{|z|}, \quad z \in \Gamma_\varepsilon. \qquad (8.5.36)$$

8.6   Appendix I

1. Continuity of the distribution function. Let us consider the Lévy process X_t from the class II. In this case the corresponding distribution $F(x,t)$ is continuous with respect to x (see Theorem 8.3.1). It follows from the representation (7.1.2), that

$$\int_{-\infty}^{\infty} e^{ixz}\, dF(x,t) = \mu(z,t). \tag{8.6.1}$$

Hence, the equality

$$\int_{0}^{x} (F(y,t) - F(-y,t))\, dy = \int_{\infty}^{\infty} \mu(\xi,t)\frac{1 - \cos(x\xi)}{\xi^2}\, d\xi. \tag{8.6.2}$$

holds [114, vol. II, Chap. XVI]. Now, we shall prove the following fact.

Theorem 8.6.1. *Let X_t be a Lévy process from the class II. Then the corresponding distribution function $F(x,t)$ is jointly continuous with respect to $t > 0$ and $x \in \mathbb{R}$.*

Proof. Using (8.6.2) and the dominated convergence theorem we deduce that the left-hand side of (8.6.2) is continuous with respect to $t > 0$ and $x \in \mathbb{R}$. Hence, the function

$$G(x,t) = F(x,t) - F(-x,t) \tag{8.6.3}$$

is continuous with respect to $t > 0$. Suppose that X_t satisfies conditions

$$A = 0, \quad 0 \in S(\nu), \quad \int_{-\infty}^{\infty} |x|\, d\nu < \infty \tag{8.6.4}$$

from item 3) of Theorem 8.2.3 and $\gamma = 0$. In this case we have either

$$F(-x,t) = 0, \quad x > 0, \quad \text{or} \quad F(x,t) = 0, \quad x > 0.$$

Since the function $G(x,t)$ is continuous with respect to $t > 0$, the function $F(x,t)$ is continuous with respect to $t > 0$ too. Next, let us consider an arbitrary Lévy process X_t from the class II. We can represent X_t in the form

$$X_t = X_t^{(1)} + X_t^{(2)}, \tag{8.6.5}$$

where Lévy processes $X_t^{(1)}$ and $X_t^{(2)}$ are independent and the process $X_t^{(2)}$ satisfies conditions (8.6.4) from Theorem 8.2.3. The processes $X_t^{(1)}$ and $X_t^{(2)}$ have the Lévy triplets (A, γ, ν_1) and $(0, 0, \nu_2)$, respectively, where

$$\nu(x) = \nu_1(x) + \nu_2(x), \quad \int_{-\infty}^{\infty} d\nu_2(x) = \infty. \tag{8.6.6}$$

The distribution functions of X_t, $X_t^{(1)}$ and $X_t^{(2)}$ are denoted by $F(x,t)$, $F_1(x,t)$ and $F_2(x,t)$, respectively. The convolution formula for Stieltjes–Fourier transform [9, Chap. 4] gives

$$F(x,t) = \int\limits_{-\infty}^{\infty} F_2(x - \xi, t - \tau) \, d_\xi F_1(\xi, \tau). \tag{8.6.7}$$

Since the process $X_t^{(2)}$ belongs to class II and satisfies conditions (8.6.4), the distribution $F_2(x,t)$ is continuous with respect to x and with respect to t. The assertion of the theorem follows from this fact and from relation (8.6.7). □

Example 8.6.2. Let us consider the case

$$\int\limits_{0}^{\infty} d\nu(x) = \infty \tag{8.6.8}$$

and explain the method of constructing Lévy measures $\nu_1(x)$ and $\nu_2(x)$. We set

$$\nu_1(x) = \nu(x) - \nu_2(x), \quad \nu_2(x) = 0 \ (x < 0), \tag{8.6.9}$$
$$d\nu_2(x) = x \, d\nu(x) \ (0 \le x \le 1), \quad d\nu_2(x) = d\nu(x) \ (x > 1). \tag{8.6.10}$$

It is easy to see that the corresponding Lévy processes $X_t^{(1)}$ and $X_t^{(2)}$ satisfy (8.6.5), are independent, and the process $X_t^{(2)}$ satisfies condition 3.

2. Non-zero points of the spectrum.

Theorem 8.6.3. *Let the following conditions be fulfilled.*

   I. *Lévy process X_t belongs to class II.*
  II. *The corresponding quasi-potential B is compact in the Banach space D_Δ and has the form (8.5.31).*
 III. *The inequality $K(x,y) > 0$ is valid, where x and y are inner points of Δ.*

Then we have:

   1. *The operator B has a point of the spectrum different from zero.*
   2. *The corresponding eigenfunction $g_1(x)$ is positive in the inner points of Δ.*
   3. *The corresponding eigenfunction $h_1(x)$ is strictly monotonic.*

Proof. Let a continuous, non-negative function $u(x)$ in the domain Δ be such that $\|u(x)\| = 1$ and

$$u(x) = 0, \quad x \in [a_k, a_k + \varepsilon_k] \cup [b_k - \varepsilon_k, b_k], \tag{8.6.11}$$
$$0 < \varepsilon_k < \frac{b_k - a_k}{2}, \quad 1 \le k \le n.$$

The function $v(x) = Bu(x)$ will be continuous in the domain Δ and $v(x) > 0$, where x are inner points of Δ. Hence, there exists such $c > 0$ that

$$v(x) \geq cu(x). \tag{8.6.12}$$

Assertion 1 of the theorem follows directly from (8.6.12) and the Kreĭn–Rutman result [45, Theorem 6.2]. The assertions 2 and 3 follow directly from condition III of the theorem. □

3. Weakly singular operators. The operator

$$Bf = \int_\Delta K(x,y)f(y)\,\mathrm{d}y \tag{8.6.13}$$

is weakly singular if the kernel $K(x,y)$ satisfies the inequality

$$|K(x,y)| \leq M|x-y|^{-r}, \quad 0 < r < 1, \quad M \equiv \mathrm{const.} \tag{8.6.14}$$

Proposition 8.6.4. *The weakly singular operator B is bounded and compact in the spaces $L^p(\Delta)$, $1 \leq p \leq \infty$.*

In particular, the weakly singular operator B is bounded and compact in the Hilbert space $L^2(\Delta)$. Repeating the arguments of Theorem 8.5.5 in the case $L^2(\Delta)$, we obtain the following assertion.

Theorem 8.6.5. *Let Lévy process X_t belong to class* II *and let the corresponding quasi-potential B be weakly singular and be strongly sectorial. Then we have*

1. $$K(x,y) = \frac{\partial}{\partial y}\Phi(x,y). \tag{8.6.15}$$

2. *The equality (8.5.10) is valid, where $c > 0$.*

Proof. Now, we consider the operator B in the Hilbert space $L^2(\Delta)$. We note that the operator B has the same eigenvalues λ_j and the same eigenvectors g_j in the spaces $L^2(\Delta)$ and D_Δ. The operator B^* has the same eigenvalues $\overline{\lambda}_j$ in the spaces $L^2(\Delta)$ and D_Δ^*. The corresponding eigenvectors $H_j(x)$ and $h_j(x)$ (in the spaces $L^2(\Delta)$ and D_Δ^*, respectively) are connected by the relation

$$H_j(x) = h_j'(x). \tag{8.6.16}$$

Recall that $W(B)$ is the numerical range of B and the contour Γ_ε is defined in the Section 8.5. The closure of the convex hull of $W(B)$ is situated in the sector (8.5.6). Hence, the estimate

$$\left\|(B - zI)^{-1}\right\| \leq \frac{M}{|z|}, \quad z \in \Gamma_\varepsilon \tag{8.6.17}$$

is valid [103]. If $r < 1/2$, then $K(0, x) \in L^2(\Delta)$. According to (8.6.17) we have

$$\left| \langle (B - zI)^{-1}1, K(0, x) \rangle \right| \leq \frac{M}{|z|}, \quad z \in \Gamma_\varepsilon. \tag{8.6.18}$$

All the conditions of Theorem 8.5.5 are fulfilled. The relation (8.5.10) is proved, if $0 < r < 1/2$. Let us consider the case when $1/2 \leq r < 1$. In this case for some $m > 0$ we obtain the relation

$$(B^*)^m K(0, x) \in L^2(\Delta). \tag{8.6.19}$$

Using relations (8.6.17), (8.6.19) and the equality

$$(B - zI)^{-1} = -\left(I + B/z + (B/z)^2 + \cdots + (B/z)^{m-1}\right) + (B - zI)^{-1}(B/z)^m, \tag{8.6.20}$$

we deduce the inequality (8.5.9). Here we have taken into account that the functions $B^k 1$ are bounded. All the conditions of Theorem 8.5.5 are fulfilled. The relation (8.5.10) is proved. Using inequality $g_1(0) > 0$ and relation (8.5.29) we obtain, that $c_1 > 0$. The theorem is proved. □

Remark 8.6.6. If X_t is the stable process, then the corresponding quasi-potential operator B is weakly singular and strongly sectorial, that is, in this case all the conditions of Theorem 8.6.5 are fulfilled (see Section 8.7 (Appendix II)).

8.7 Appendix II (stable processes)

1. In the famous article [30] by M. Kac, a number of examples demonstrate interconnections between probability theory and problems of the theory of integral and differential equations. In particular, symmetric stable processes are considered in [30]. The study of these processes comes down to the solution of integro-differential equations of a special type [30, 33]. However, as M. Kac writes, the solution of the obtained equations presents great analytical difficulties. M. Kac himself solved the equation, which appears in the study of Cauchy processes.

In the author's work [87] all the symmmetric stable processes were investigated and answers to all the questions of M. Kac, referring to stable processes, were obtained. In the present chapter, the results of Chapter 2 are used for the study of non-symmetric case. We use the analytic apparatus of the theory of integral equations with a difference kernel. The integral operators considered in this chapter also play an important role in the contact theory of elasticity [102] and in the extremal problems of interpolation theory.

2. Let us describe the principal definitions.

Let $X_1, X_2, \ldots$ be mutually independent random variables with the same law of distribution $F(x)$. The distribution $F(x)$ is called strictly stable if the random variable

$$X = \frac{X_1 + X_2 + \cdots + X_n}{C_n}, \tag{8.7.1}$$

$$C_n = n^{1/\alpha} \tag{8.7.2}$$

is also distributed according to the law $F(x)$. The number α $(0 < \alpha \le 2)$ is called a characteristic exponent of the distribution. When $\alpha \ne 1$ the relations (8.7.1), (8.7.2) are equivalent to the equality

$$E\left[e^{i\xi X}\right] = \exp\left(-\lambda|\xi|^\alpha\left(1 - i\beta(\operatorname{sgn}\xi)\tan(\pi\alpha/2)\right)\right) \tag{8.7.3}$$

where $|\beta| \le 1$, $\lambda > 0$. If $\beta = 0$ the random variable X is symmetric. If $|\beta| = 1$, then X is called an entirely asymmetric random variable. When $\alpha = 2$ we obtain a normally distributed random variable. The homogeneous process $X(\tau)$ $(X(0) = 0)$ with independent increments is called a stable process if

$$E\left[e^{i\xi X(\tau)}\right] = \exp\left(-\tau|\xi|^\alpha\left(1 - i\beta(\operatorname{sgn}\xi)\tan(\pi\alpha/2)\right)\right). \tag{8.7.4}$$

Stable processes are a natural generalization of Wiener processes. Comparing (8.7.3) and (8.7.4) we see that when τ is fixed, $X(\tau)$ is a strictly stable random variable with the parameters α, β, $\lambda = \tau$. If $\alpha \ne 1$ then the corresponding operator S has the form (see Chapter 7):

$$S_\alpha f = \frac{\sin(\pi\alpha/2)}{\pi}\Gamma(\alpha - 1)\int\limits_{-a}^{a}\left(1 - \beta\operatorname{sgn}(x - \xi)\right)|x - \xi|^{1-\alpha}f(\xi)\,d\xi. \tag{8.7.5}$$

3. The explicit form of the quasi-potential operator B, solution of the Kac problems. First, we consider the case

$$|\beta| < 1. \tag{8.7.6}$$

We introduce the functions

$$\mathcal{L}_1(x) = \mathcal{D}(x + a)^{-\rho}(a - x)^{\rho - \mu}, \tag{8.7.7}$$

$$\mathcal{L}_2(x) = \frac{(2\rho - \mu - 1)a + x}{1 - \mu}\mathcal{L}_1(x) \tag{8.7.8}$$

where $\mu = 2 - \alpha$ and parameters ρ and $\mathcal{D}$ are defined by the relations:

$$\tan(\pi\rho) = \frac{(1 - \beta)\sin(\pi\mu)}{(1 + \beta) + (1 - \beta)\cos(\pi\mu)}, \quad 0 < \rho < 1, \tag{8.7.9}$$

$$\mathcal{D} = \frac{\sin(\pi\rho)}{\sin(\pi\alpha/2)(1 - \beta)\Gamma(\alpha - 1)}. \tag{8.7.10}$$

Relation (8.7.9) can be written in the form

$$\sin(\pi\rho) = \frac{1-\beta}{1+\beta}\sin\left(\pi(\mu-\rho)\right). \tag{8.7.11}$$

It follows that

$$0 < \mu - \rho < 1. \tag{8.7.12}$$

Taking into account the transition from the segment $[0, \omega]$ to the segment $[-a, a]$, from Theorem 3.1.1 we deduce the following assertion.

Theorem 8.7.1. *Let $|\beta| < 1$, $\alpha \neq 1$. Then the equalities*

$$S_\alpha \mathcal{L}_k = x^{k-1}, \quad k = 1, 2 \tag{8.7.13}$$

are valid.

Setting

$$r = \int_{-a}^{a} \mathcal{L}_1(x)\,dx \tag{8.7.14}$$

we introduce the functions

$$q(x,y) = \frac{\mathcal{L}_1(-y)\mathcal{L}_2(x) - \mathcal{L}_2(-y)\mathcal{L}_1(x)}{r}, \tag{8.7.15}$$

$$\Phi(x,y) = \frac{1}{2}\int_{x+y}^{2a-|x-y|} q\left(\frac{s+x-y}{2}, \frac{s-x+y}{2}\right) ds. \tag{8.7.16}$$

The corresponding quasi-potential operator B has the form (Section 2.5)

$$Bf = \int_{-a}^{a} \Phi(x,y)\,dy. \tag{8.7.17}$$

Remark 8.7.2. If we know $\mathcal{L}_1(x)$ and $\mathcal{L}_2(x)$ we can construct the corresponding operator B in the explicit form.

By formula (8.7.15) we have

$$q(x,y) = C_\alpha(a^2 - y^2)^{-\rho}(a^2 - x^2)^{-\rho}\left((a+y)(a-x)\right)^{2\rho-\mu}(x+y), \tag{8.7.18}$$

where

$$C_\alpha = \frac{(2a)^{\mu-1}\sin(\pi\rho)}{\left(\sin(\pi\alpha/2)\right)(1-\beta)\Gamma(1-\rho)\Gamma(1+\rho-\mu)}. \tag{8.7.19}$$

Substituting (8.7.18) into (8.7.16) and setting $z = a^2 - (s^2/4) + (x - y)^2/4$, we deduce the relation

$$\Phi_\alpha(x, y) = C_\alpha \int\limits_{a|x-y|}^{a^2-xy} \left(z^2 - a^2(x - y)^2\right)^{-\rho} \left(z - a(x - y)\right)^{2\rho-\mu} dz. \qquad (8.7.20)$$

Making one more change $z = |x - y|ua$ we come to the equality

$$\Phi_\alpha(x, y) = C_\alpha a^{1-\mu}|x - y|^{1-\mu} \int\limits_{1}^{(a^2-xy)/(a|x-y|)} \left(u^2 - 1\right)^{-\rho} \left(u - \operatorname{sgn}(x - y)\right)^{2\rho-\mu} du.$$
$$\qquad (8.7.21)$$

From formula (8.7.19) it follows that

$$C_\alpha > 0. \qquad (8.7.22)$$

Using relations (8.7.20) and (8.7.21) we see the validity of the following theorem.

Theorem 8.7.3. *Let the inequality $|\beta| < 1$ be valid. Then we have*

1. *The function $\Phi_\alpha(x, y)$ is non-negative and*

$$\Phi_\alpha(x, \pm a) = 0, \quad -a < x < a. \qquad (8.7.23)$$

2. *The function $\Phi_\alpha(x, y)$ is continuous when $1 < \alpha < 2$.*
3. *The function $\Phi_\alpha(x, y)$ in the case $0 < \alpha < 1$ has a discontinuity only when $x = y$ and*

$$\Phi_\alpha(x, y) \le M|x - y|^{\alpha-1}. \qquad (8.7.24)$$

Now we turn to the case $\beta = 1$, $1 < \alpha < 2$. In this case we have:

$$S_\alpha f = \frac{2}{\pi} \sin\left(\frac{\pi\alpha}{2}\right) \Gamma(\alpha - 1) \int\limits_{x}^{a} f(y)(y - x)^{1-\alpha} dy, \qquad (8.7.25)$$

$$\mathcal{L}_1(x) = -\cos\left(\frac{\pi\alpha}{2}\right) \frac{(a - x)^{\alpha-2}}{\Gamma(\alpha - 1)}, \qquad (8.7.26)$$

$$\mathcal{L}_2(x) = \mathcal{L}_1(x)\frac{(\alpha - 3)a + x}{\alpha - 1}. \qquad (8.7.27)$$

Again, by means of the formula (8.7.17) we introduce the operator B_α. According to (8.7.15), (8.7.16) and (8.7.26), (8.7.27) we have

$$\Phi_\alpha(x, y) = \cos\left(\frac{\pi\alpha}{2}\right) (2a)^{1-\alpha} \frac{(2a|y - x|)^{\alpha-1} - (a - x)^{\alpha-1}(a + y)^{\alpha-1}}{\Gamma(\alpha)}, \qquad (8.7.28)$$

where $1 < \alpha < 2$ and $y > x$. $\qquad \Longleftarrow$

If $1 < \alpha < 2$ and $y < x$, then

$$\Phi_\alpha(x,y) = -\cos\left(\frac{\pi\alpha}{2}\right)(2a)^{1-\alpha}\frac{(a-x)^{\alpha-1}(a+y)^{\alpha-1}}{\Gamma(\alpha-1)}, \tag{8.7.29}$$

Let us consider the case $\beta = -1$, $1 < \alpha < 2$. In this case we have:

$$S_\alpha f = \frac{2}{\pi}\sin\left(\frac{\pi\alpha}{2}\right)\Gamma(\alpha-1)\int\limits_{-a}^{x} f(y)(x-y)^{1-\alpha}\,dy, \tag{8.7.30}$$

$$\mathcal{L}_1(x) = -\cos\left(\frac{\pi\alpha}{2}\right)\frac{(a+x)^{\alpha-2}}{\Gamma(\alpha-1)}, \tag{8.7.31}$$

$$\mathcal{L}_2(x) = \mathcal{L}_1(x)\frac{(1-\alpha)a+x}{\alpha-1}. \tag{8.7.32}$$

Again, by means of formula (8.7.17) we introduce the operator B_α. The kernel of the operator B has the form:

$$\Phi_\alpha(x,y) = \cos\left(\frac{\pi\alpha}{2}\right)(2a)^{1-\alpha}\frac{(2a|x-y|)^{\alpha-1}-(a-y)^{\alpha-1}(a+x)^{\alpha-1}}{\Gamma(\alpha)}, \tag{8.7.33}$$

where $1 < \alpha < 2$ and $y < x$. If $1 < \alpha < 2$ and $y > x$, then

$$\Phi_\alpha(x,y) = -\cos\left(\frac{\pi\alpha}{2}\right)\frac{(2a)^{1-\alpha}(a-y)^{\alpha-1}(a+x)^{\alpha-1}}{\Gamma(\alpha-1)}. \tag{8.7.34}$$

Taking into account (8.7.28), (8.7.29) and (8.7.33), (8.7.34) we obtain the following assertion.

Proposition 8.7.4. *If $\beta = \pm 1$, $1 < \alpha < 2$, then the kernel $\Phi_\alpha(x,y)$ of the operator B is bounded.*

Now, we consider the case $\beta = \pm 1$, $0 < \alpha < 1$.

Proposition 8.7.5.

1. If $\beta = 1$, $0 < \alpha < 1$, then formulas (8.7.28) and (8.7.29) are valid.
2. If $\beta = -1$, $0 < \alpha < 1$, then formulas (8.7.33) and (8.7.34) are valid.

Proof. Let $\beta = 1$. Formulas (8.7.25), (8.7.28) and (8.7.29) are well defined under the conditions $0 < \operatorname{Re}\alpha < 2$, $\alpha \neq 1$. The relation $B_\alpha L_\Delta = I$ is fulfilled, when $1 < \operatorname{Re}\alpha < 2$. Using analytic continuation we obtain that the relation $B_\alpha L_\Delta = I$ holds, when $0 < \operatorname{Re}\alpha < 1$. The case $\beta = -1$ can be investigated in the same way. The proposition is proved. $\qquad\square$

Taking into account the inequality

$$(2a|x-y|)^{\alpha-1} \geq (a-y)^{\alpha-1}(a+x)^{\alpha-1}, \quad 0 < \alpha < 1, \tag{8.7.35}$$

we obtain the following statement.

Theorem 8.7.6. *If $\beta = \pm 1$, $0 < \alpha < 1$, then the corresponding quasi-potential operator B_α is weakly singular.*

The Lévy measure in the case of $\alpha = 1$, $\beta = 0$ has form (8.4.41) with

$$\eta_1 = \eta_2 = 0, \quad C_1 = C_2 = 1, \quad \Delta = [-a, a].$$

Using formulas (7.2.2), (7.2.3) and (7.2.12), (7.2.13) we obtain:

$$k(x) = -\ln \frac{2|x|}{b}. \tag{8.7.36}$$

According to formulas (3.2.1) and (3.2.13) we have

$$\Phi_1(x,t) = -\frac{1}{\pi^2} \ln \frac{2a|x-t|}{\left(\sqrt{(x+a)(a-t)} + \sqrt{(t+a)(a-x)}\right)^2}. \tag{8.7.37}$$

The following assertion is valid.

Proposition 8.7.7. *If $\alpha = 1$, $\beta = 0$, then the corresponding quasi-potential operator B_1 belongs to the Hilbert–Schmidt class and $\Phi_1(0,t) \in L^2(-a, a)$.*

Let us consider three cases:

   I. $0 < \alpha < 2$, $\alpha \neq 1$, $-1 < \beta < 1$,

  II. $0 < \alpha < 2$, $\alpha \neq 1$, $\beta = \pm 1$,

 III. $\alpha = 1$, $\beta = 0$.

Theorem 8.7.8. *Let X_t be stable Lévy process and let the corresponding parameters α and β satisfy one of the conditions I-III. Then the corresponding quasi-potential operator B satisfies all the conditions of Theorem 8.5.5.*

Proof. The quasi-potential operator B is strongly sectorial (see Corollary 8.4.23 and Corollary 8.4.25).

The quasi-potential operator B is either weakly singular or belongs to the Hilbert–Schmidt class (see Theorem 8.7.3, Propositions 8.7.4, 8.7.5 and 8.7.7). This proves the theorem. $\qquad\square$

Chapter 9

Triangular Factorization and Cauchy Type Lévy Processes

9.1 Introduction

1. Let us introduce the main notions of the triangular factorization (see [82, 88, 92, 96]). First, we define the orthogonal projectors P_ξ in the Hilbert space $L^2(a,b)$ ($-\infty \leq a < b \leq \infty$):

$$P_\xi f = \begin{cases} f(x), & a \leq x < \xi, \\ 0, & \xi < x \leq b, \end{cases} \qquad \text{where} \quad f(x) \in L^2(a,b).$$

Definition 9.1.1. A bounded operator S_- on $L^2(a,b)$ is called lower triangular if for every ξ the relations

$$S_- Q_\xi = Q_\xi S_- Q_\xi, \tag{9.1.1}$$

where $Q_\xi = I - P_\xi$, are valid. The operator S_-^* is called upper triangular.

Definition 9.1.2. A bounded, positive definite and invertible operator S on $L^2(a,b)$ is said to admit left (right) triangular factorization if it can be represented in the form

$$S = S_- S_-^* \quad (S = S_-^* S_-), \tag{9.1.2}$$

where S_- and S_-^{-1} are lower triangular, bounded operators.

In the paper [96] we formulated necessary and sufficient conditions under which a positive definite operator S admits triangular factorization. The factorizing operator S_-^{-1} was constructed in the explicit form. We proved that a wide class of the bounded, positive definite and invertible operators admits triangular factorization.

2. *Now, we do not suppose that the corresponding operator S is invertible.*
We use a factorization method for integral equations of the first kind:

$$Sf = \int_0^\omega k(x,t)f(t)\,\mathrm{d}t = \varphi(x), \tag{9.1.3}$$

where the operator S is positive definite. We assume that there exists such a
function $h(x)$, that

$$|k(x,t)| \le h(x-t), \quad \int_{-\omega}^\omega h(x)\,\mathrm{d}x < \infty. \tag{9.1.4}$$

Introduce the notations

$$S_\xi = P_\xi SP_\xi, \quad \langle f,g \rangle_\xi = \int_0^\xi f(x)\overline{g(x)}\,\mathrm{d}x. \tag{9.1.5}$$

Further, we assume that the following conditions are fulfilled:

1) There exists a function $F_0(x) \in L^q(0,\omega)$, $q \ge 2$, such that the solution $v(\xi,x)$
   of the integral equation

$$S_\xi\{v(\xi,x)\} = F_0(x), \quad 0 \le x \le \xi \le \omega, \tag{9.1.6}$$

   belongs to $L^p(0,\xi)$, where $\dfrac{1}{p} + \dfrac{1}{q} = 1$.

2) The function

$$M(\xi) = \langle F_0(x), v(\xi,x) \rangle_\xi \tag{9.1.7}$$

   is absolutely continuous and almost everywhere

$$M'(\xi) \ne 0. \tag{9.1.8}$$

In terms of $v(\xi,x)$ and $M(\xi)$, we construct the operator V:

$$Vf = \left(R(x)\right)^{-1}\frac{\mathrm{d}}{\mathrm{d}x}\int_0^x \overline{v(x,t)}f(t)\,\mathrm{d}t, \tag{9.1.9}$$

where $R(x) = \sqrt{M'(x)}$. Under certain natural conditions we prove the following
statement.

Proposition 9.1.3. *The function*

$$f(x) = V^*V\varphi \tag{9.1.10}$$

is the solution of the equation (9.1.3).

Formula (9.1.10) can be rewritten in the triangular factorization form:

$$S^{-1} = V^*V. \tag{9.1.11}$$

We note that the operators S^{-1}, V and V^* are unbounded.

M.G. Kreĭn formulated without proof the result (see [21, Chap. 4, formula (8.24)]), which coincides with Proposition 9.1.3 in the particular case $F_0 = 1$. We apply Proposition 9.1.3 to the case

$$k(x,t) = k(|x - t|) \tag{9.1.12}$$

and consider in detail the following examples:

$$k(x) = \ln \frac{b}{2|x|}, \quad b > 0; \tag{9.1.13}$$

$$k(x) = \ln \frac{\sin(b/2)}{2 \sin(|x|/2)}, \quad b > 0; \tag{9.1.14}$$

$$k(x) = \ln \frac{\sinh(b/2)}{2 \sinh(|x|/2)}, \quad b > 0; \tag{9.1.15}$$

$$k(x) = |x|^{1-\alpha}, \quad 1 < \alpha < 2. \tag{9.1.16}$$

We note that examples (9.1.13)–(9.1.16) are related to the important cases of the Lévy processes.

9.2 Integral equations of the first kind and triangular factorization

1. It follows from (9.1.7)–(9.1.9) that

$$V F_0 = R(x), \quad R(x) \in L^2(0, \omega). \tag{9.2.1}$$

By D_V and by R_V we denote the domain of definition and the range of the operator V, respectively. We assume that for some $q \geq 2$ the relations

$$D_V \subset L^q(0, \omega), \quad R_V \subset L^2(0, \omega) \tag{9.2.2}$$

are valid.

Proposition 9.2.1. *Let D_V be dense in the space $L^q(0, \omega)$. Then*

$$V^* P_\xi V F_0 = v(\xi, t), \tag{9.2.3}$$

where $v(\xi, t) = 0$, $\xi < t \leq \omega$.

Proof. Equality (9.2.3) follows directly from the relation

$$\langle VF_0, Vg\rangle_\xi = \int_0^\xi v(\xi, t)\overline{g(t)}\,dt. \tag{9.2.4}$$

$\square$

Equality (9.1.6) can be written in the form

$$P_\xi Sv(\xi, t) = P_\xi F_0. \tag{9.2.5}$$

Further, we suppose that the following conditions are fulfilled:
The operator V^{-1} is bounded in the space $L^2(0, \omega)$ and has the triangular property

$$P_\xi V^{-1} P_\xi V P_\xi = P_\xi. \tag{9.2.6}$$

From (9.2.3) and (9.2.6) we obtain the relation

$$P_\xi F_0 = P_\xi V^{-1}(V^{-1})^* v(\xi, x). \tag{9.2.7}$$

Relations (9.2.5) and (9.2.7) imply that

$$\left\langle V^{-1}(V^{-1})^* v(\xi, x), v(\eta, x)\right\rangle = \langle Sv(\xi, x), v(\eta, x)\rangle. \tag{9.2.8}$$

If the system $v(\xi, t)$ $(0 \le \xi \le \omega)$ is complete in the space $L^p(0, \omega)$ then in view of relation (9.2.8) we have

$$S = V^{-1}(V^{-1})^*. \tag{9.2.9}$$

So the following result is proved:

Theorem 9.2.2. *Let the operator V be constructed by relations (9.1.6)–(9.1.10). Let the following conditions be fulfilled:*

1. *The set D_V is dense in the space $L^q(0, \omega)$ and (9.2.2) holds.*
2. *The operator V^{-1} is bounded in the space $L^2(0, \omega)$ and the triangular property (9.2.6) is valid.*
3. *The set of functions $v(\xi, x)$ $(0 \le \xi \le \omega)$ is complete in the space $L^p(0, \omega)$.*

Then the equality (9.2.9) holds.

We note that according to (9.2.2) the relations

$$D_{V^{-1}} = R_V \subset L^2(0, \omega), \quad R_{V^{-1}} = D_V \subset L^q(0, \omega) \tag{9.2.10}$$

are valid. Relations (9.2.10) imply that

$$D_{(V^{-1})^*} \subset L^p(0, \omega) \quad \left(\frac{1}{p} + \frac{1}{q} = 1\right), \qquad R_{(V^{-1})^*} \subset L^2(0, \omega). \tag{9.2.11}$$

By R_S we denote the set of such functions $\varphi(x)$ that

$$\varphi \in D_V, \quad V\varphi \in D_{V^*}. \tag{9.2.12}$$

Now we can formulate the main theorem of this section.

Theorem 9.2.3. *Let the conditions of Theorem 9.2.2 be fulfilled. Then the function*

$$f(x) = V^*V\varphi(x), \quad \varphi(x) \in R_S \tag{9.2.13}$$

is the solution of the equation

$$Sf = \varphi. \tag{9.2.14}$$

If $g(x)/R(x)$ is a differentiable function then according to (9.1.9) the operator V^* has the form

$$V^*g = v(\omega, x)\frac{g(\omega)}{R(\omega)} - \int_x^\omega v(t, x)\frac{d}{dt}\frac{g(t)}{R(t)}\,dt. \tag{9.2.15}$$

Remark 9.2.4. If we put (9.2.15) into (9.2.13) we obtain the Kreĭn result [21, Chap. 4, Formula (8.24)]. M.G. Kreĭn proves this result only for integral equations of the second type (the case where the corresponding operator S is invertible) and announced without proof that this result is valid for integral equations of the first kind too (the case, when the corresponding operator S is not invertible).

2. Further, we consider an important class of integral equations

$$Sf = \int_{-a}^a k(|x - t|)f(t)\,dt = \varphi(x). \tag{9.2.16}$$

This class of operators S is used in a number of applied problems [1, 21, 24, 110] and in the Lévy processes theory [97]. Let us introduce the special solutions $v_m(\xi, x)$ $(m = 1, 2)$ of equation (9.2.16) when $a = \xi$.

$$Sv_m(\xi, x) = \int_{-\xi}^\xi k(|x - t|)v_m(\xi, t)\,dt = x^{m-1}, \quad m = 1, 2. \tag{9.2.17}$$

We assume that

$$v_m(\xi, x) \in L^p(-\xi, \xi), \quad 1 < p \le 2. \tag{9.2.18}$$

It is easy to see that

$$v_1(\xi, x) = v_1(\xi, -x), \quad v_2(\xi, x) = -v_2(\xi, -x). \tag{9.2.19}$$

Relations (9.2.17) and (9.2.19) imply

$$\int_0^\xi \Big(k(|x - t|) + k(x + t)\Big)v_1(\xi, t)\,dt = 1, \quad 0 \le x \le \xi, \tag{9.2.20}$$

$$\int_0^\xi \Big(k(|x - t|) - k(x + t)\Big)v_2(\xi, t)\,dt = x, \quad 0 \le x \le \xi. \tag{9.2.21}$$

9.3   Triangular factorization, examples

1. We consider an integral equation with a kernel of the power type (see Section 3.1), where

$$\beta = 0, \quad 1 < \alpha < 2. \tag{9.3.1}$$

In this case we have

$$k(x,t) = k(x-t), \quad k(x) = |x|^{1-\alpha}. \tag{9.3.2}$$

If $F_0 = 1$ then the relation

$$v(\xi,t) = \frac{1}{\pi} \sin \frac{\pi\alpha}{2} \left(t(\xi - t) \right)^{\frac{\alpha}{2}-1} \tag{9.3.3}$$

holds (Section 3.1). It follows from (9.1.7) and (9.3.3) that

$$M(\xi) = c\xi^{\alpha-1}, \quad c = \frac{\Gamma^2(\alpha/2)}{\pi\Gamma(\alpha)} \sin \frac{\pi\alpha}{2}. \tag{9.3.4}$$

Relation (9.3.4) implies that

$$R(\xi) = \sqrt{M'(\xi)} = \sqrt{c(\alpha-1)}\xi^{\alpha/2-1}. \tag{9.3.5}$$

Using (9.1.9), (9.3.3) and (9.3.5) we construct the corresponding operator V in the explicit form

$$Vf = c_1 x^{1-\alpha/2} \frac{\mathrm{d}}{\mathrm{d}x} \int_0^x \left(t(x-t) \right)^{\alpha/2-1} f(t)\, \mathrm{d}t, \tag{9.3.6}$$

where

$$c_1 = \frac{\sin(\pi\alpha/2)}{\pi\sqrt{c(\alpha-1)}}. \tag{9.3.7}$$

Let us calculate the expression Vx^m:

$$Vx^m = c_1 \frac{\Gamma(\alpha/2)\Gamma(\alpha/2+m)}{\Gamma(\alpha+m-1)} x^{\alpha/2+m-1}, \quad m \geq 0. \tag{9.3.8}$$

Proposition 9.3.1. *The operator V^{-1} has the form*

$$V^{-1}f = c_2 x^{1-\alpha/2} \int_0^x \left(t(x-t) \right)^{-\alpha/2} t^{\alpha-1} f(t)\, \mathrm{d}t, \tag{9.3.9}$$

where

$$c_2 = \sqrt{c(\alpha-1)}. \tag{9.3.10}$$

Proof. Using well-known formulas for gamma and beta functions we calculate the expression $V^{-1}x^n$:

$$V^{-1}x^n = c_2 \frac{\Gamma(1-\alpha/2)\Gamma(\alpha/2+n)}{\Gamma(n+1)} x^{-\alpha/2+n+1}, \quad n \geq 0. \tag{9.3.11}$$

In view of (9.3.7) and (9.3.11) we have

$$VV^{-1}x^n = x^n, \quad V^{-1}Vx^m = x^m. \tag{9.3.12}$$

The proposition is proved. $\square$

In case (9.3.2) relation (9.2.15) has the form

$$V^*g = c_1 \left(\omega^{1-\alpha/2}\big(x(\omega-x)\big)^{\alpha/2-1} g(\omega) - \int\limits_x^\omega \big(x(t-x)\big)^{\alpha/2-1} \frac{d}{dt}\big(g(t)t^{1-\alpha/2}\big)\, dt \right). \tag{9.3.13}$$

We put

$$G_m(x) = V^*x^{\alpha/2+m-1}. \tag{9.3.14}$$

It follows from (9.3.13) and (9.3.14) that

$$|G_m(x)| \leq C\big(x(\omega-x)\big)^{\alpha/2-1}(1+x^{m-1}). \tag{9.3.15}$$

Proposition 9.3.2. *If*

$$m > 1 - \frac{\alpha}{2}, \tag{9.3.16}$$

then $x^m \in D_V$ and $Vx^m \in D_V^*$.

Proof. According to relations (9.3.15) and (9.3.16) there exists such $p > 1$ that $G_m(x) \in L^p(0,\omega)$. The proposition is proved. $\square$

It is easy to see that all conditions of Theorem 9.2.2 are fulfilled and we have the following result.

Theorem 9.3.3. *Let relations (9.3.2) and (9.3.16) be valid. Then the function $f_m(x) = V^*Vx^m$ is the solution of the corresponding integral equation*

$$Sf_m(x) = x^m, \quad m > 1 - \frac{\alpha}{2}. \tag{9.3.17}$$

The next assertion follows directly from Theorem 2.2.1 and Theorem 9.3.3.

Proposition 9.3.4. *Let relations (9.3.2) and (9.3.16) be valid. Then there exists the solution $f(x) \in L^p(0,\omega)$ $(p > 1)$ of the corresponding integral equation (9.2.16), where $\varphi(x) = cx^m + \varphi_1(x)$, $\varphi_1(x) \in W_p^{(2)}$.*

2. We consider the case (Section 3.2), where

$$k(x,t) = k(x-t), \quad k(x) = \ln\left(\frac{b}{2}|x|\right), \quad b > 0, \quad b \neq \frac{\omega}{2}. \tag{9.3.18}$$

If $F_0 = 1$ then the relations

$$v(\xi,t) = \frac{M(\xi)}{\pi}(t(\xi-t))^{-1/2}, \quad M(\xi) = \int_0^\xi v(\xi,t)\,dt = \left(\ln\left(\frac{2b}{\xi}\right)\right)^{-1} \tag{9.3.19}$$

are valid (see Section 3.2). Relation (9.3.19) implies that

$$R(\xi) = \sqrt{M'(\xi)} = \left(\ln\left(\frac{2b}{\xi}\right)\right)^{-1}\xi^{-1/2}. \tag{9.3.20}$$

Using (9.1.9), (9.3.19) and (9.3.20) we construct the corresponding operator V in the explicit form

$$Vf = \left(\ln\left(\frac{2b}{x}\right)\right)x^{1/2}\frac{d}{dx}\int_0^x\left(\ln\left(\frac{2b}{t}\right)\right)^{-1}(t(x-t))^{-1/2}f(t)\,dt. \tag{9.3.21}$$

Let us introduce the operator V_1:

$$V_1 f = x^{1/2}\frac{d}{dx}\int_0^x (t(x-t))^{-1/2}f(t)\,dt. \tag{9.3.22}$$

We have

$$V_1 x^m = \sqrt{\pi}\frac{\Gamma(1/2+m)}{\Gamma(m)}x^{m-1/2}, \quad m \geq 0. \tag{9.3.23}$$

Proposition 9.3.5. *The operator V_1^{-1} has the form*

$$V_1^{-1}f = \frac{1}{\pi}x^{1/2}\int_0^x (t(x-t))^{-1/2}f(t)\,dt. \tag{9.3.24}$$

Proof. Using well-known formulas for gamma and beta functions we calculate the expression $V_1^{-1}x^{m-1/2}$:

$$V_1^{-1}x^{m-1/2} = \frac{1}{\sqrt{\pi}}\frac{\Gamma(m)}{\Gamma(m+1/2)}x^m, \quad m \geq 0. \tag{9.3.25}$$

In view of (9.3.7) and (9.3.11) we have

$$V_1\left(V_1^{-1}\right)x^m = x^m, \quad \left(V_1^{-1}\right)V_1 x^m = x^m. \tag{9.3.26}$$

The proposition is proved. $\square$

Using relation (9.3.21) and Proposition 9.3.5 we obtain the following statement.

Proposition 9.3.6. *The operator V^{-1} has the form*

$$V^{-1}f = \frac{1}{\pi} x^{1/2} \ln \left(\frac{2b}{x} \right) \int_0^x \left(\ln \left(\frac{2b}{t} \right) \right)^{-1} (t(x-t))^{-1/2} f(t) \, dt. \qquad (9.3.27)$$

9.4 New examples of the Lévy processes

In this section we introduce and study two new Lévy processes. These processes are not stable but in a certain sense they are similar to the Cauchy stable processes.

Definition 9.4.1. We say that the Lévy process X_t is a Cauchy type process, if the equalities $A = 0$ and $\gamma = 0$ hold, and the corresponding Lévy measure $\nu(x)$ is such that $\lim_{x \to 0} x^2 \nu'(x) > 0$.

We note that the Lévy measure in the case of Cauchy process is such that

$$\nu'(x) = \frac{1}{x^2 \pi}.$$

Example 9.4.2. Let the operator S_Δ have the form

$$S_\Delta f = \int_{-a}^a \ln \frac{\sin (b/2)}{2 \sin (|x-t|/2)} f(t) \, dt, \quad 0 < a < b \le \pi. \qquad (9.4.1)$$

The corresponding kernel $k(x)$ has the form

$$k(x) = \ln \frac{\sin (b/2)}{2 \sin (|x|/2)}. \qquad (9.4.2)$$

Hence, we have

$$k'(x) = -\frac{1}{2} \cot(|x|/2) \operatorname{sgn} x, \quad k''(x) = \frac{1}{4 \sin^2 (x/2)}. \qquad (9.4.3)$$

According to relations (7.2.2), (7.2.3) and (7.2.12), (7.2.13) the Lévy measure $\nu(x)$ and the kernel $k(x)$ are connected by the equality

$$\nu'(x) = k''(x). \qquad (9.4.4)$$

From (9.4.2)–(9.4.4) we obtain the following statement.

Proposition 9.4.3. *If the operator S_Δ has the form (9.4.2), then the corresponding Lévy process X_t is a Cauchy type process.*

We introduce the solutions $\mathcal{L}_k(x)$, $k = 1, 2$, of the equations

$$S_\Delta \mathcal{L}_k = x^{k-1}, \quad k = 1, 2. \tag{9.4.5}$$

Theorem 9.4.4. *In case of equation* (9.4.1) *the solutions* $\mathcal{L}_k$, $k = 1, 2$, *are defined by the formulas*

$$\mathcal{L}_1(x) = \frac{M(a)}{2\pi} \left(\cos^2 \frac{x}{2} - \cos^2 \frac{a}{2} \right)^{-1/2} \cos \frac{x}{2}, \tag{9.4.6}$$

$$M(a) = \left(\ln \frac{\sin(b/2)}{\sin(a/2)} \right)^{-1}, \tag{9.4.7}$$

$$\mathcal{L}_2(x) = \frac{\sin(x/2)}{\pi} \left(\cos^2 \frac{x}{2} - \cos^2 \frac{a}{2} \right)^{-1/2}. \tag{9.4.8}$$

Remark 9.4.5. Formula (9.4.6) was obtained by M.G. Kreĭn [21, Chap. 4]. Formula (9.4.8) follows from (9.2.13) and (9.4.6).

Theorem 9.4.6. *Let the operator* S_Δ *have the form* (9.4.1). *Then the following assertions are valid:*

1. *The corresponding quasi-potential B is defined by the relation*

$$Bf = \int_{-a}^{a} \Phi(x, y) f(y) \, \mathrm{d}y, \tag{9.4.9}$$

   where $\Phi(x, y)$ has the form (2.5.12) *(see formula* (2.5.4) *and Theorem* 2.5.2).
2. *The operator $T = S^{-1}$ is defined by formula* (2.2.4).

Using formulas (9.4.6)–(9.4.8), Theorem 9.4.6 and Theorem 8.5.5 we obtain the following assertion.

Proposition 9.4.7. *In case* (9.4.1) *the quasi-potential operator B is compact in the space $L^2(\Delta)$.*

Example 9.4.8. Let the operator S_Δ have the form

$$S_\Delta f = \int_{-a}^{a} \ln \frac{\sinh(b/2)}{2 \sinh \left(|x - t|/2 \right)} f(t) \, \mathrm{d}t, \quad 0 < a < b. \tag{9.4.10}$$

The corresponding kernel $k(x)$ has the form

$$k(x) = \ln \frac{\sinh(b/2)}{2 \sinh \left(|x|/2 \right)}. \tag{9.4.11}$$

Hence, we have

$$k'(x) = -\frac{1}{2} \coth\left(|x|/2\right) \operatorname{sgn} x, \quad k''(x) = \frac{1}{4 \sinh^2\left(x/2\right)}. \tag{9.4.12}$$

According to relations (7.2.2), (7.2.3) and (7.2.12), (7.2.13) the Lévy measure $\nu(x)$ and the kernel $k(x)$ are connected by the equality

$$\nu'(x) = k''(x). \tag{9.4.13}$$

Relations (9.4.11)–(9.4.13) yeld the following statement.

Proposition 9.4.9. *The corresponding Lévy process X_t is a Cauchy type process.*

We introduce the solutions $\mathcal{L}_k(x,a)$, $k = 1,2$, of the equations

$$S_\Delta \mathcal{L}_k = x^{k-1}, \quad k = 1,2. \tag{9.4.14}$$

Theorem 9.4.10. *In the case of equation (9.4.10) the solutions $\mathcal{L}_k$, $k = 1,2$, are defined by the formulas*

$$\mathcal{L}_1(a,x) = \frac{M(a)}{2\pi}\left(\cosh^2\frac{a}{2} - \cosh^2\frac{x}{2}\right)^{-1/2}\cosh\frac{x}{2}, \tag{9.4.15}$$

$$M(a) = \left(\ln\frac{\sinh\left(b/2\right)}{\sinh\left(a/2\right)}\right)^{-1}, \tag{9.4.16}$$

$$\mathcal{L}_2(a,x) = \frac{\sinh\left(x/2\right)}{\pi}\left(\cosh^2\frac{a}{2} - \cosh^2\frac{x}{2}\right)^{-1/2}. \tag{9.4.17}$$

Remark 9.4.11. Formula (9.4.15) was obtained by M.G. Kreĭn [21, Chap. 4]. Formula (9.4.17) follows from (9.2.13) and (9.4.15).

Theorem 9.4.12. *Let the operator S_Δ have the form (9.4.10). Then the following assertions are valid:*

1. *The corresponding quasi-potential B is defined by the relation*

$$Bf = \int_{-a}^{a} \Phi(x,y)f(y)\,dy, \tag{9.4.18}$$

   where $\Phi(x,y)$ has the form (2.5.12) (see formula (2.5.4) and Theorem 2.5.2).
2. *The operator $T = S^{-1}$ is defined by formula (2.2.4).*

Using formulas (9.4.15)–(9.4.17), Theorem 9.4.12 and Theorem 8.5.5 we obtain the following assertion.

Proposition 9.4.13. *In case (9.4.10) the quasi-potential operator B is compact in the space $L^2(\Delta)$.*

9.5   Two-sided estimation of the smallest eigenvalue of the operator $(-L_\Delta)$

1. Let λ_1 and φ_1 denote the largest eigenvalue and the corresponding eigenfunction of the quasi-potential operator B. The value $\mu_1 = 1/\lambda_1$ characterizes how fast $p(t, \Delta)$ converges to zero when $t \to +\infty$. The knowledge of μ_1 is also essential for solving some interpolation problems (see [97, Chap. 3]).

We consider the case when $\Delta = [-a, a]$, $a > 0$. The operator L_Δ has the form

$$L_\Delta f = \frac{d}{dx} S_\Delta \frac{d}{dx} f, \quad f(-a) = f(a) = 0. \tag{9.5.1}$$

We assume that the following conditions are fulfilled.

1. The operator S_Δ is positive definite and has the form

$$S_\Delta f = \int\limits_{-a}^{a} k(x - t) f(t)\, dt, \tag{9.5.2}$$

where the kernel $k(x)$ is real, $k(x) \in L(-2a, 2a)$ and $k(x) = k(-x)$.

2. There exists a function $f_0(x) \in L(-a, a)$ such that

$$S_\Delta f_0(x) = x. \tag{9.5.3}$$

It follows from (9.5.2) and (9.5.3) that

$$f_0(x) = -f_0(-x). \tag{9.5.4}$$

Hence, the function

$$g_0(x) = \int\limits_{x}^{a} f_0(t)\, dt \tag{9.5.5}$$

satisfies the relations

$$g_0(a) = g_0(-a) = 0. \tag{9.5.6}$$

Using equality

$$\langle -L_\Delta g_0(x), g_0(x) \rangle = \langle S_\Delta f_0(x), f_0(x) \rangle \tag{9.5.7}$$

we deduce the following assertion.

Proposition 9.5.1. *Let conditions* (9.5.2) *and* (9.5.3) *be fulfilled. Then*

$$\mu_1 \leq \int_{-a}^{a} x f_0(x)\, dx \Big/ \int_{-a}^{a} g_0^2(x)\, dx. \tag{9.5.8}$$

2. Any function $\psi(x) \geq 0$, $\psi(x) \in L^2(-a, a)$, satisfies the relations

$$\lambda_1 = \frac{\langle B\varphi_1, \psi \rangle}{\langle \varphi_1, \psi \rangle} = \frac{\langle \varphi_1, (B\psi/\psi)\psi \rangle}{\langle \varphi_1, \psi \rangle}. \tag{9.5.9}$$

According to the properties of the quasi-potential B we have that $B\psi \geq 0$. Hence, relations (9.5.9) imply

$$\lambda_1 \leq \max_{-a \leq x \leq a} \left\{ \frac{B\psi(x)}{\psi(x)} \right\}. \tag{9.5.10}$$

Thus, we have obtained a formula for a lower bound of $\mu_1 = \dfrac{1}{\lambda_1}$:

$$\mu_1 \geq \left(\max_{-a \leq x \leq a} \left\{ \frac{B\psi(x)}{\psi(x)} \right\} \right)^{-1}. \tag{9.5.11}$$

Remark 9.5.2. Relations (9.5.8) and (9.5.11) were used in the book [97, p. 51], and estimates for μ_1 for the case (9.1.16) were obtained there.

3. Here we shall derive new estimates for μ_1. Recall that $\Delta = [-a, a]$ and so, in view of (9.2.17), the function $f_0(x) = v_2(a, x)$ satisfies (9.5.3). Then, according to Proposition 9.5.1, (9.5.8) holds. Rewrite (9.5.8) in the form

$$\mu_1 \leq 2M_2(a) \Big/ \int_{-a}^{a} g_0^2(x)\, dx \quad \text{for} \quad 2M_2(a) = \int_{-a}^{a} x f_0(x)\, dx. \tag{9.5.12}$$

Now we estimate the integral $\displaystyle\int_{-a}^{a} g_0^2(x)\, dx$ from below:

$$2a^{-1} \left(\int_{0}^{a} |g_0(x)|\, dx \right)^2 \leq 2 \int_{0}^{a} g_0^2(x)\, dx = \int_{-a}^{a} g_0^2(x)\, dx. \tag{9.5.13}$$

Using (9.5.5) we obtain

$$\int_{0}^{a} g_0(x)\, dx = M_2(a). \tag{9.5.14}$$

The next statement follows from relations (9.5.8) and (9.5.12)–(9.5.14).

Proposition 9.5.3. *If $v_2(a, x) \geq 0$ $(0 \leq x \leq a)$, then $g_0(x) \geq 0$ $(-a \leq x \leq a)$, and*

$$\mu_1 \leq \frac{a}{M_2(a)}. \tag{9.5.15}$$

Now, we use the inequality (9.5.11), where $\psi(x) = g_0(x)$. It is easy to see that

$$-L_\Delta g_0(x) = 1, \quad -BL_\Delta g_0(x) = g_0(x). \tag{9.5.16}$$

If $f_0(x) \geq 0$ $(0 \leq x \leq a)$, then $g_0(x) \geq 0$ $(-a \leq x \leq a)$ and

$$\max_{-a \leq x \leq a} \{g_0(x)\} = g_0(0). \tag{9.5.17}$$

According to (9.5.11), (9.5.16) and (9.5.17) we have the following statement.

Proposition 9.5.4. *If $v_2(a, x) \geq 0$ $(0 \leq x \leq a)$, then $g_0(x) \geq 0$ $(-a \leq x \leq a)$, and*

$$\mu_1 \geq \frac{1}{g_0(0)}. \tag{9.5.18}$$

Example 9.5.5. Let the kernel $k(x)$ be defined by formula (9.4.2).

We use formulas (9.5.8) and (9.5.11). It follows from (9.4.6)–(9.4.8) that

$$v_1(\xi, x) = \frac{M_1(\xi)}{2\pi} \left(\cos^2 \frac{x}{2} - \cos^2 \frac{\xi}{2} \right)^{-1/2} \cos \frac{x}{2}, \tag{9.5.19}$$

$$M_1(\xi) = \left(\ln \frac{\sin(b/2)}{\sin(\xi/2)} \right)^{-1}, \tag{9.5.20}$$

$$v_2(\xi, x) = \frac{\sin(x/2)}{\pi} \left(\cos^2 \frac{x}{2} - \cos^2 \frac{\xi}{2} \right)^{-1/2}. \tag{9.5.21}$$

According to (9.4.8) and (9.5.20) we have

$$M_2'(\xi) = 2\frac{\sin(\xi/2)}{\cos(\xi/2)}, \quad M_2(\xi) = -2 \ln \left(\cos \frac{\xi}{2} \right). \tag{9.5.22}$$

In view of (9.5.21) the equality

$$g_0(x) = \int_x^a v_2(a, t)\, dt = \frac{2}{\pi} \ln \frac{\cos(x/2) + \sqrt{\cos^2(x/2) - \cos^2(a/2)}}{\cos(a/2)} \tag{9.5.23}$$

holds. Hence, the relations

$$g_0(x) \leq g_0(0) = \frac{2}{\pi} \ln \frac{1 + \sin(a/2)}{\cos(a/2)} \tag{9.5.24}$$

are valid.

Propositions 9.5.1 and 9.5.3 with relations (9.5.22) and (9.5.24) imply the following assertion.

Proposition 9.5.6. *Let the kernel $k(x)$ be defined by formula (9.4.2). Then*

$$\frac{\pi}{2}\left(\ln\frac{1+\sin(a/2)}{\cos(a/2)}\right)^{-1} \le \mu_1(a) \le \frac{a}{2\left|\ln\left(\cos(a/2)\right)\right|}. \tag{9.5.25}$$

Remark 9.5.7. When parameter a is small we have

$$\frac{\pi}{2}\left(\ln\frac{1+\sin(a/2)}{\cos(a/2)}\right)^{-1} \sim \frac{\pi}{a}, \quad \frac{a}{2\left|\ln\left(\cos(a/2)\right)\right|} \sim \frac{4}{a}. \tag{9.5.26}$$

Example 9.5.8. Let the kernel $k(x)$ be defined by formula (9.4.11). We use formulas (9.5.8) and (9.5.11). It follows from (9.4.15)–(9.4.17) that

$$v_1(\xi,x) = \frac{M_1(\xi)}{2\pi}\left(\cosh^2\frac{\xi}{2} - \cosh^2\frac{x}{2}\right)^{-1/2}\cosh\frac{x}{2}, \tag{9.5.27}$$

$$M_1(\xi) = \left(\ln\frac{\sinh(b/2)}{\sinh(\xi/2)}\right)^{-1}, \tag{9.5.28}$$

$$v_2(\xi,x) = \frac{\sinh(x/2)}{\pi}\left(\cosh^2\frac{\xi}{2} - \cosh^2\frac{a}{2}\right)^{-1/2}. \tag{9.5.29}$$

According to (9.4.17) and (9.5.29) we have

$$M_2'(\xi) = 2\frac{\sinh(\xi/2)}{\cosh(\xi/2)}, \quad M_2(\xi) = 2\ln\left(\cosh(\xi/2)\right). \tag{9.5.30}$$

In view of (9.5.29) the equality

$$g_0(x) = \int_x^a v_2(a,t)\,dt = \frac{2}{\pi}\left(\frac{\pi}{2} - \arcsin\frac{\cosh(x/2)}{\cosh(\xi/2)}\right) \tag{9.5.31}$$

holds. Hence, the relations

$$g_0(x) \le g_0(0) = \frac{2}{\pi}\left(\frac{\pi}{2} - \arcsin\frac{1}{\cosh(a/2)}\right) \tag{9.5.32}$$

are valid. Propositions 9.5.1 and 9.5.3 with relations (9.5.29) and (9.5.31) imply the following assertion.

Proposition 9.5.9. *Let the kernel $k(x)$ be defined by formula (9.4.11). Then*

$$\frac{1}{1 - (2/\pi)\arcsin\left(1/\cosh(a/2)\right)} \le \mu_1(a) \le \frac{a}{2\ln\cosh(a/2)}. \tag{9.5.33}$$

Remark 9.5.10. When parameter a is small, we have

$$\frac{1}{1 - (2/\pi)\arcsin\left(1/\cosh(a/2)\right)} \sim \frac{2}{a}, \quad \frac{a}{2\ln\cosh(a/2)} \sim \frac{2}{a}. \tag{9.5.34}$$

Remark 9.5.11. For large values of a we have

$$\frac{1}{1 - (2/\pi)\arcsin\left(1/\cosh(a/2)\right)} \sim 1, \quad \frac{a}{2\ln\cosh(a/2)} \sim 1. \tag{9.5.35}$$

Proposition 9.5.12. *Let the kernel $k(x)$ be defined by formula (9.4.11). Then*

$$\lim_{a \to 0} a\mu_1(a) = 2, \quad \lim_{a \to \infty} \mu_1(a) = 1. \tag{9.5.36}$$

The following result is obtained in the book [97].

Proposition 9.5.13. *Let the kernel $k(x)$ be defined by formula (9.1.16). Then*

$$a^{-\alpha}\Gamma(\alpha + 1) \leq \mu_1(a) \leq \frac{a^{-\alpha}\Gamma\left(1 + \dfrac{\alpha}{2}\right)\Gamma\left(\alpha + \dfrac{3}{2}\right)}{\Gamma\left(\dfrac{3 + \alpha}{2}\right)}, \tag{9.5.37}$$

where $\Gamma(z)$ is the Euler gamma function.

9.6 Proof of Kreĭn's formulas

In this section we prove some of M.G. Kreĭn's formulas [21, Chap. 4, Section 8], which were announced without proof. We show that

   I. these formulas are valid for wider classes than announced by M.G. Kreĭn;

  II. these formulas are factorization formulas.

We consider the positive definite operator

$$S_\Delta f(t) = \int\limits_{-a}^{a} k(x - t)f(t)\,\mathrm{d}t = \varphi(t), \tag{9.6.1}$$

where the kernel $k(x)$ is real, $k(x) \in L(-2a, 2a)$, and $k(x) = k(-x)$. If the function $\varphi(t)$ is even, then the corresponding solution $f(t)$ is even too and we have

$$H_1\varphi = \int\limits_{0}^{a} \left(k\big(|x - t|\big) + k(x + t)\right)f(t)\,\mathrm{d}t = \varphi(t), \quad 0 \leq x \leq a. \tag{9.6.2}$$

If the function $\varphi(t)$ is odd, then the corresponding solution $f(t)$ is odd too and we have

$$H_2 f = \int_0^a \Big(k(|x - t|) - k(x + t)\Big) f(t)\, dt = \varphi(t), \quad 0 \leq x \leq a. \tag{9.6.3}$$

Introduce the operators

$$V_k f = \big(M_k'(x)\big)^{-1/2} \frac{d}{dx} \int_0^x v_k(x, t) f(t)\, dt, \quad k = 1, 2, \tag{9.6.4}$$

where $v_k(x, t)$ and $M_k(x)$ are defined by formulas (9.2.20) and (9.2.21). Assume that $F_0 = 1$ when $k = 1$ and $F_0 = x$ when $k = 2$. Let conditions of Theorem 9.2.2 be fulfilled. Then we have

$$H_k^{-1} = V_k^* V_k. \tag{9.6.5}$$

M.G. Kreĭn (see [21, Chap. 4, formulas (8.24) and (8.25)]) proved this result for integral equations of the second kind

$$S_\Delta f(t) = \mu f + \int_{-a}^a k(x - t) f(t)\, dt = \varphi(t), \quad \mu > 0, \tag{9.6.6}$$

and announced without proof that this result is also valid under some conditions for integral equations of the first kind. That is, he announced that this result is valid for the case $\mu = 0$. Kreĭn's approach is based on special triangular factorization [21, Chap. 4], where triangular operators have a less general triangular form. Our approach to the problems of triangular factorization is based on the general triangular form (see [82, 88, 96]). Using the relation

$$V_1^* g = \frac{v_1(a, x) g(a)}{\sqrt{M_1'(a)}} - \int_x^a v_1(t, x) \frac{d}{dt} \frac{g(t)}{\sqrt{M'(t)}}\, dt, \tag{9.6.7}$$

we see that formula (9.6.5) for $k = 1$ coincides with Kreĭn's formula [21, Chap. 4, formula (8.24)].

Now, we prove an important formula, which describes the connection between $v_1(t, x)$ and $v_2(t, x)$. We consider a positive definite operator

$$S_\xi f(t) = \int_0^\xi k(x - t) f(t)\, dt, \tag{9.6.8}$$

where the kernel $k(x)$ is real, $k(x) \in L(-\omega, \omega)$ $(0 \leq \xi \leq \omega)$, and $k(x) = k(-x)$. Introduce the function

$$s(x) = \int_0^x k(u)\, du, \quad -\omega \leq x \leq \omega, \tag{9.6.9}$$

and the solutions $N_1(\xi, x)$, $\mathcal{L}_k(\xi, x)$ $(k = 1, 2)$ of the equations

$$S_\xi N_1(\xi, x) = s(x), \quad S_\xi \mathcal{L}_k(\xi, x) = x^{k-1}, \quad 0 \le x \le \xi, \quad k = 1, 2. \qquad (9.6.10)$$

Let us rewrite (1.2.3), for $m = 1$, in the form

$$\mathcal{L}_2(\xi, x) - C(\xi)\mathcal{L}_1(\xi, x) = M_1(\xi)N_1(\xi, x) - \int_x^\xi \mathcal{L}_1(\xi, t)\, dt, \qquad (9.6.11)$$

where

$$M_1(\xi) = \int_0^\xi \mathcal{L}_1(\xi, t)\, dt, \quad C(\xi) = \int_0^\xi \mathcal{L}_1(\xi, t)s(t)\, dt. \qquad (9.6.12)$$

Using (9.6.10) and (9.6.11) we obtain the following assertion.

Proposition 9.6.1. *Assume that the solutions* $\mathcal{L}_k(\xi, x)$ $(k = 1, 2)$ *exist and belong to the space* $L^p(0, \xi)$ $(1 < p)$ *and that*

$$M_1'(\xi) > 0, \quad \int_0^\xi \left| \mathcal{L}_1(\xi, t)s(t) \right| dt < \infty. \qquad (9.6.13)$$

Then the solution $N_k(\xi, x)$ *exists and belongs to the space* $L^p(0, \xi)$ $(1 < p)$.

Remark 9.6.2. If the operator S_ξ is strongly positive definite, then the inequality $M_1'(\xi) > 0$ follows from relation $M_1(\xi) = \langle S_\xi \mathcal{L}_1(\xi, x), \mathcal{L}_1(\xi, x) \rangle > 0$.

Next, we need the following result (see either [95, Formula (49)] or [97, Formula (7.2.33)]).

Proposition 9.6.3. *Let the conditions of Proposition 9.6.1 be fulfilled. Then*

$$\int_0^\xi N_1(\xi, t)\, dt = \int_0^\xi \mathcal{L}_1(\xi, t)s(t)\, dt = \frac{1}{2}\xi. \qquad (9.6.14)$$

It follows from (9.6.13) and (9.6.11) that

$$\langle \mathcal{L}_2(\xi, t), \mathbb{1} \rangle = M_1(\xi)\xi - \langle \mathcal{L}_1(\xi, t), t \rangle. \qquad (9.6.15)$$

Taking into account the equality

$$\langle \mathcal{L}_2(\xi, t), \mathbb{1} \rangle = \langle \mathcal{L}_1(\xi, t), t \rangle \qquad (9.6.16)$$

we obtain

$$\langle \mathcal{L}_1(\xi, t), t \rangle = \frac{\xi}{2} M_1(\xi). \qquad (9.6.17)$$

We consider the case where $F_0 = \mathbb{1}$. In this case we have $v_1(t,x) = \mathcal{L}_1(t,x)$. Formulas (9.1.9) and (9.6.17) imply that

$$g_0(x) = Vx = \frac{M_1(x) + xM_1'(x)}{2\sqrt{M_1'(x)}}. \qquad (9.6.18)$$

According to (9.1.11), (9.2.15) and (9.6.16), the relation

$$v_2(\xi,x) = \frac{v_1(\xi,x)g_0(\xi)}{R(\xi)} - \int_x^\xi v_1(t,x)\frac{d}{dt}\frac{g_0(t)}{R(t)}\,dt, \qquad (9.6.19)$$

where

$$R(x) = \sqrt{M_1'(x)}, \qquad (9.6.20)$$

is valid.

Chapter 10

Lévy Processes with Summable Lévy Measures, Long Time Behavior

10.1 Introduction

1. In Chapter 7 we proved that the infinitesimal generators L for Lévy processes X_t can be represented in a convolution-type form. In the case of a non-summable Lévy measure we constructed a quasi-potential operator B and investigated the long time behavior of X_t (Chapters 8 and 9). $\Longleftarrow$ +par

In the present chapter we consider a Lévy process X_t with a summable Lévy measure. In this case the quasi-potential operator B has a form, which is essentially different from the corresponding form in the case of a Lévy process with the non-summable Lévy measure. We use this new form and study again the long time behavior of X_t.

2. In the present chapter we consider Lévy processes from class I (see Chapter 7, Definition 7.1.3). Without loss of generality we assume that

$$\gamma - \int_{|y|<1} y\nu(\,\mathrm{d}y) = 0. \tag{10.1.1}$$

According to (7.1.7), the generator L of the corresponding process X_t can be represented in the form

$$Lf = \int_{-\infty}^{\infty} \big(f(x+y) - f(x)\big)\nu(\,\mathrm{d}y). \tag{10.1.2}$$

Remark 10.1.1. If condition (10.1.1) is valid, then class I coincides with the class of the compound Poisson processes.

As in the case of class II we use the convolution representation of the generator L. To do it we introduce the functions

$$\mu_-(x) = \int_{-\infty}^{x} \nu(dx), \quad x < 0, \tag{10.1.3}$$

$$\mu_+(x) = -\int_{x}^{\infty} \nu(dx), \quad x > 0, \tag{10.1.4}$$

where the functions $\mu_-(x)$ and $\mu_+(x)$ are monotonically increasing and right continuous on the half-axis $(-\infty, 0]$ and $[0, \infty)$, respectively. We note that

$$\mu_-(x) \to 0, \quad x \to -\infty; \qquad \mu_+(x) \to 0, \quad x \to +\infty, \tag{10.1.5}$$

$$\mu_-(x) \ge 0, \quad x < 0; \qquad \mu_+(x) \le 0, \quad x > 0. \tag{10.1.6}$$

Now, we define the functions

$$k_-(x) = \int_{-b}^{x} \mu_-(t)\, dt, \quad -b \le x < 0, \quad b > 0, \tag{10.1.7}$$

$$k_+(x) = -\int_{x}^{b} \mu_+(t)\, dt, \quad 0 < x \le +b. \tag{10.1.8}$$

Proposition 10.1.2. *Let the Lévy process X_t belong to class* I *and let the condition* (10.1.9) *be fulfilled. Then formula* (10.1.2) *can be written in the convolution form*

$$Lf = \frac{d}{dx} S \frac{d}{dx} f, \tag{10.1.9}$$

where

$$Sf = \int_{-\infty}^{\infty} k(y - x) f(y)\, dy, \qquad k(x) = \begin{cases} k_+(x), & x > 0, \\ k_-(x), & x < 0. \end{cases} \tag{10.1.10}$$

Using (10.1.7)–(10.1.10) we obtain the following assertion.

Proposition 10.1.3. *Let conditions of Proposition* 10.1.2 *be fulfilled. Then relation* (10.1.9) *takes the form*

$$Lf = -\Omega f + \int_{-\infty}^{\infty} f(y)\, d_y \nu(y - x), \qquad \Omega = \int_{-\infty}^{\infty} d\nu(y). \tag{10.1.11}$$

Proof. It follows from (10.1.7)–(10.1.10) that

$$Lf = -\int_{-\infty}^{x} \mu_-(y-x)f'(y)\,dy - \int_{x}^{\infty} \mu_+(y-x)f'(y)\,dy. \qquad (10.1.12)$$

Integrating by parts (10.1.12) we have

$$Lf = -\big(\mu_-(0) - \mu_+(0)\big)f + \int_{-\infty}^{\infty} f(y)\,d_y\nu(y-x). \qquad (10.1.13)$$

The proposition is proved. $\qquad\square$

In formulas (10.1.11) and (10.1.13) we use the equality $d\nu(x) = \nu(dx)$.

Definition 10.1.4.

1. We say that Lévy process X_t belongs to the class I_c if the corresponding Lévy measure $\nu(y)$ is summable and continuous.
2. We say that Lévy process X_t belongs to the class I_d if the corresponding Lévy measure $\nu(y)$ is summable and discrete.

It is easy to see that the following assertion is valid.

Proposition 10.1.5. *If Lévy process X_t belongs to the class* I, *then X_t can be represented in the form*

$$X_t = X_t^{(1)} + X_t^{(2)}, \qquad (10.1.14)$$

where $X_t^{(1)} \in I_c$ and $X_t^{(2)} \in I_d$.

The main part of the present chapter is dedicated to the study of the Lévy processes from the class I_c.

As before, we denote by $p(t, \Delta)$ the probability that a sample of the process $X_t \in I_c$ remains inside the domain Δ for $0 \le \tau \le t$ (ruin problem). Using representation (10.1.9) we derive a new formula for $p(t, \Delta)$. This formula allows us to obtain a long time behavior of $p(t, \Delta)$. Namely, we prove the asymptotic formula

$$p(t, \Delta) = e^{-t/\lambda_1}\big(c_1 + o(1)\big), \quad c_1 > 0, \quad \lambda_1 > 0, \quad t \to +\infty. \qquad (10.1.15)$$

Let T_Δ be the time during which X_t remains in the domain Δ before it leaves the domain Δ for the first time. It is easy to see that

$$p(t, \Delta) = P\big(T_\Delta > t\big). \qquad (10.1.16)$$

An essential role in our theory is played by the operator

$$L_\Delta f = -\Omega f + \int_{\Delta} f(y)\,d_y\nu(y-x), \qquad (10.1.17)$$

which is generated by the operator L (see (10.1.11)). We note that λ_1 in formula (10.1.15) is the largest eigenvalue of $-L_\Delta^{-1}$.

Definition 10.1.6. The measure $\nu(y)$ is unimodal with mode 0 if $\nu(y)$ is concave for $y < 0$ and convex for $y > 0$.

The unimodality and its properties were actively investigated (see[99]). In the present chapter we derive a new important property of the unimodal measure.

Proposition 10.1.7. *If Lévy measure $\nu(y)$ is continuous, summable and unimodal with mode 0, then the operator L_Δ^{-1} has the form*

$$L_\Delta^{-1} = -\frac{1}{\Omega}(I + T_1), \qquad (10.1.18)$$

where the operator T_1 is compact in the space of continuous functions.

In the last part of this chapter we investigate the operator L_Δ when $X_t \in I_d$.

10.2   Quasi-potential

1. Recall that Δ denotes the set of segments $[a_k, b_k]$, where

$$a_1 < b_1 < a_2 < b_2 < \cdots < a_n < b_n, \quad 1 \le k \le n.$$

By D_Δ we denote the space of the continuous functions $g(x)$ on the domain Δ. The norm in D_Δ is defined by the relation $\|f\| = \sup_{x \in \Delta} |f(x)|$. The space D_Δ^0 is defined by the relations

$$f(x) \in D_\Delta \quad \text{and} \quad f(a_k) = f(b_k) = 0, \quad 1 \le k \le n.$$

We introduce the operator P_Δ by the relation $P_\Delta f(x) = \begin{cases} f(x), & x \in \Delta, \\ 0, & x \notin \Delta. \end{cases}$

Definition 10.2.1. The operator

$$L_\Delta = P_\Delta L P_\Delta = \frac{\mathrm{d}}{\mathrm{d}x} S_\Delta \frac{\mathrm{d}}{\mathrm{d}x}, \text{ where } S_\Delta = P_\Delta S P_\Delta, \qquad (10.2.1)$$

is called a truncated generator.

$$\left(\text{We use here the equality } P_\Delta \frac{\mathrm{d}}{\mathrm{d}x} = \frac{\mathrm{d}}{\mathrm{d}x} P_\Delta, \, x \in \Delta. \right)$$

Definition 10.2.2. The operator B with the domain of definition D_Δ is called a quasi-potential if the relation

$$-BL_\Delta g = g, \quad g \in C_\Delta \qquad (10.2.2)$$

is valid.

According to Proposition 10.1.3, the operator L_Δ has the form

$$L_\Delta f = -\Omega f + \int_\Delta f(y)\, d_y \nu(y-x), \quad f(x) \in D^0_\Delta. \tag{10.2.3}$$

We introduce the operator

$$T f = \int_\Delta f(y)\, d_y \nu(y-x), \quad f(x) \in D^0_\Delta. \tag{10.2.4}$$

Equality (10.2.4) implies the following statement.

Theorem 10.2.3. *Let the Lévy process X_t belong to class* I *and let condition* (10.1.9) *be fulfilled. Then the operator T acts from D_Δ into D_Δ and*

$$\|T\| = \sup \int_\Delta d_y \nu(y-x) \leq \Omega, \quad x \in \Delta. \tag{10.2.5}$$

Further, we suppose in addition that

$$\|T\| < \Omega. \tag{10.2.6}$$

Remark 10.2.4. Let the support of the Lévy measure ν be unbounded. Then in view of (10.2.5) the inequality (10.2.6) holds.

Now, we consider separately the case where the Lévy process X_t belongs to the class I_c.

Theorem 10.2.5. *Let X_t belong to class I_c and let the condition* (10.2.6) *be true. Then the operator L_Δ^{-1} exists and has the form*

$$L_\Delta^{-1} = -B = -\frac{1}{\Omega}(I + T_1), \tag{10.2.7}$$

where the operator T_1, which acts in the space D_Δ, is bounded and defined by the formula

$$T_1 f = \int_\Delta f(y)\, d_y \Phi(x,y). \tag{10.2.8}$$

The function $\Phi(x,y)$ is continuous with respect to x and y and monotonically increasing with respect to y.

Proof. The assertion of the theorem follows from the relation

$$T_1 = \frac{T}{\Omega} + \left(\frac{T}{\Omega}\right)^2 + \cdots . \tag{10.2.9}$$

$\square$

10.3   Quasi-potential and compactness of T and T_1

1. In this section we consider the following problem:

Under which conditions are the operators T and T_1 compact in the space D_Δ?

We recall that the operators T and T_1 are defined by (10.2.4) and (10.2.8) respectively. J. Radon [61] proved the following theorem.

Theorem 10.3.1. *The operator T_1, which is defined by formula* (10.2.8), *is compact in the space D_Δ if and only if*

$$\lim_{x \to \xi} \left\| \Phi(x, y) - \Phi(\xi, y) \right\|_V = 0, \quad x, y, \xi \in \Delta. \tag{10.3.1}$$

Hence, we have the following statement.

Proposition 10.3.2. *If measure $\nu(y)$ is summable and has continuous derivative, then the corresponding operators T and T_1 are compact in the space D_Δ.*

Proof. Relation (10.2.4) takes the form:

$$Tf = \int_\Delta f(y)\nu'(y - x)\,dy, \quad x \in \Delta. \tag{10.3.2}$$

The conditions of Radon's theorem are fulfilled, that is, the operator T is compact. In view of (10.2.4), (10.2.5) and (10.2.8) the operator T_1 is compact as well. The proposition is proved.   □

2. Let us consider the important case, where the Lévy measure is unimodal (see Definition 10.1.6). We shall need the following properties of convex and concave functions.

Proposition 10.3.3. *Let the points x_p $(p = 1, 2, 3, 4)$ be such that $x_1 < x_2 \le x_3 < x_4$.*

1. *If a function $f(x)$ is convex then*

$$\frac{f(x_2) - f(x_1)}{x_2 - x_1} \ge \frac{f(x_4) - f(x_3)}{x_4 - x_3}. \tag{10.3.3}$$

2. *If a function $f(x)$ is concave then*

$$\frac{f(x_2) - f(x_1)}{x_2 - x_1} \le \frac{f(x_4) - f(x_3)}{x_4 - x_3}. \tag{10.3.4}$$

Corollary 10.3.4. *The assertions 1 and 2 of Proposition 10.3.3 are valid if*

$$x_1 < x_3 \le x_2 < x_4 \quad and \quad x_2 - x_1 = x_4 - x_3.$$

Proof. Let us consider the convex case and choose such an integer n that $\dfrac{\ell}{n} < \ell_1$, where $\ell = x_2 - x_1$ and $\ell_1 = x_3 - x_1$. It follows from (10.3.3) that

$$
\sum_{k=1}^{n} \left(f\left(x_1 + \frac{k\ell}{n}\right) - f\left(x_1 + \frac{(k-1)\ell}{n}\right) \right)
$$
$$
\geq \sum_{k=1}^{n} \left(f\left(x_3 + \frac{k\ell}{n}\right) - f\left(x_3 + \frac{(k-1)\ell}{n}\right) \right). \tag{10.3.5}
$$

Hence, for the convex case the corollary is proved. In the same way the corollary can be proved for the concave case. $\qquad\square$

Theorem 10.3.5. *If a measure $\nu(\,dx)$ on R is unimodal with mode 0, then the corresponding operators T and T_1 are compact in the space D_Δ.*

Proof. Let us consider the case where $\Delta = [c, d]$ and $c = y_0 < y_1 < \cdots < y_n = d$. We introduce the variation

$$
V_n(x, \xi) = \sum_{k=1}^{n} \left| \big(\mu(y_k - x) - \mu(y_{k-1} - x)\big) - \big(\mu(y_k - \xi) - \mu(y_{k-1} - \xi)\big) \right|, \tag{10.3.6}
$$

where $c \leq \xi < x \leq d$. Without loss of generality we assume that

$$
\max \left| y_k - y_{k-1} \right| < x - \xi, \quad 1 \leq k \leq n. \tag{10.3.7}
$$

By y_N we denote such a point that

$$
y_N - x \geq 0, \quad y_{N-1} - x \leq 0. \tag{10.3.8}
$$

We represent equality (10.3.6) in the form

$$
V_n(x, \xi) = \sum_{k=1}^{N-2} |b_k| + \sum_{k=N-1}^{N} |b_k| + \sum_{N+1}^{n} |b_k|, \tag{10.3.9}
$$

where

$$
b_k = \Big(\mu\big(y_k - x\big) - \mu\big(y_{k-1} - x\big)\Big) - \Big(\mu\big(y_k - \xi\big) - \mu\big(y_{k-1} - \xi\big)\Big). \tag{10.3.10}
$$

Proposition 10.3.3 implies that

$$
\Big(\mu\big(y_k - x\big) - \mu\big(y_{k-1} - x\big)\Big) - \Big(\mu\big(y_k - \xi\big) - \mu\big(y_{k-1} - \xi\big)\Big) \geq 0, \quad k \geq N,
$$
$$
\tag{10.3.11}
$$
$$
\Big(\mu\big(y_k - x\big) - \mu\big(y_{k-1} - x\big)\Big) - \Big(\mu\big(y_k - \xi\big) - \mu\big(y_{k-1} - \xi\big)\Big) \leq 0, \quad k \leq N - 2.
$$
$$
\tag{10.3.12}
$$

It follows from (10.3.9)–(10.3.12) that

$$V_n(x,\xi) = -\sum_{k=1}^{N-2} b_k + \sum_{k=N-1}^{N} |b_k| + \sum_{N+1}^{n} b_k. \tag{10.3.13}$$

Hence, we have

$$V_n(x,\xi) \le D_0 + 2(D_{N-2} + D_{N-1} + D_N) + D_n, \tag{10.3.14}$$

where $D_k = |\mu(y_k - x) - \mu(y_k - \xi)|$. The function $\mu(y)$ is continuous. Therefore

$$\sup\{V_n(x,\xi)\} \to 0, \quad x \to \xi. \tag{10.3.15}$$

It follows from the last relation and from Radon's theorem that the assertion of the theorem is valid in the case where $\Delta = [c,d]$. Then the theorem also is valid in the case of an arbitrary domain Δ. $\qquad\square$

Remark 10.3.6. Theorem 10.3.5 is valid in the case where the measure $\nu(\,dx)$ is n-modal $(1 \le n < \infty)$.

Remark 10.3.7. The unimodality of the Lévy measure $\nu(\,dx)$ is closely connected with the unimodality of the probability distribution $F(x,t)$ of the corresponding Lévy process (see [99, Section 52]).

It is easy to obtain the following assertion.

Proposition 10.3.8. *Let condition* (10.2.6) *be fulfilled. Then the spectrum of the operator B belongs to the right half-plane. If in addition the operator T is compact, then the eigenvalues λ_j of B are such that*

$$\lambda_j \to \frac{1}{\Omega}, \quad j \to \infty. \tag{10.3.16}$$

10.4   The probability that Lévy process (class I_c) remains within the given domain

1. Recall that the definition of Lévy processes of class I_c is given in Section 10.1. Further, we assume that the following conditions are fulfilled.

Condition 10.4.1.

1. *The Lévy process X_t belongs to class I_c.*
2. *The relation* (10.1.9) *is valid.*
3. *The relation* (8.2.3) *is valid.*

We denote by $F_0(x,t)$ the distribution function of Lévy process X_t, that is,

$$F_0(x,t) = P(X_t \leq x). \qquad (10.4.1)$$

We need the following statement [99, Remark 27.3].

Theorem 10.4.2. *If the Conditions 10.4.1 are fulfilled, then the distribution function $F_0(x,t)$ is continuous with respect to x for $x \neq 0$.*

Let us investigate the behavior of $F_0(x,t)$ at the point $x = 0$.

Proposition 10.4.3. *If the Conditions 10.4.1 are fulfilled, then the relation*

$$F_0(+0,t) - F_0(-0,t) = e^{-t\Omega}, \quad \Omega = \int_{-\infty}^{\infty} \nu(\,dx), \qquad (10.4.2)$$

is valid. Here, by definition, we have

$$F(0,t) = \frac{F_0(+0,t) + F_0(-0,t)}{2}. \qquad (10.4.3)$$

Proof. In our case equality (7.1.3) takes the form

$$\lambda(z) = -\int_{-\infty}^{\infty} (e^{ixz} - 1)\nu(\,dx) = \Omega - \omega(z), \qquad (10.4.4)$$

where

$$\omega(z) = \int_{-\infty}^{\infty} e^{ixz}\nu(\,dx). \qquad (10.4.5)$$

Using the inverted Fourier–Stieltjes transform we have

$$F_0(x,t) - F_0(0,t) = \frac{1}{2\pi} e^{-t\Omega} \int_{-\infty}^{\infty} \frac{e^{-ixz} - 1}{-iz} \sum_{k=0}^{\infty} \frac{(t\omega(z))^k}{k!}\,dz. \qquad (10.4.6)$$

In view of (10.4.4) the relation

$$\mu(x) - \mu(0) = \frac{1}{2\pi} \int_{-\infty}^{\infty} \frac{e^{-ixz} - 1}{-iz}\omega(z)\,dz \qquad (10.4.7)$$

holds. The Lévy process X_t belongs to the class I_c. Then the measure $\nu(\,dx)$ is continuous. Hence, formula (10.4.7) implies, that

$$\int_{-\infty}^{\infty} \frac{e^{-ixz} - 1}{-iz}\omega(z)\,dz \bigg|_{x=0} = 0. \qquad (10.4.8)$$

In the same way we can prove the formulas

$$\int_{-\infty}^{\infty} \frac{e^{-ixz} - 1}{-iz} \omega^k(z) \, dz \bigg|_{x=0} = 0, \quad k \geq 1. \tag{10.4.9}$$

It follows from (10.4.5) and (10.4.9) that

$$F_0(+0, t) - F_0(-0, t) = e^{-t\Omega} \lim_{x \to +0} \frac{1}{\pi} \int_{-\infty}^{\infty} \frac{e^{-ixz} - 1}{-iz} \, dz = e^{-t\Omega}. \tag{10.4.10}$$

The proposition is proved. □

We introduce the sequence of functions

$$F_{n+1}(x, t) = \int_0^t \int_{-\infty}^{\infty} F_0(x - \xi, t - \tau) V(\xi) \, d_\xi F_n(\xi, \tau) \, d\tau, \quad n \geq 0, \tag{10.4.11}$$

where the function $V(x)$ is defined by the relation

$$V(x) = \begin{cases} 1, & \text{when } x \notin \Delta, \\ 0, & \text{when } x \in \Delta. \end{cases}$$

In the right-hand side of (10.4.11) we use Stieltjes integration. According to (10.4.11) the function $F_1(x, t)$ is continuous with respect to x, for $x \neq 0$. The point $x = 0$ we shall consider independently.

Theorem 10.4.4. *Let Conditions* 10.4.1 *be fulfilled. Then the functions* $F_n(x, t)$, $n > 0$, *are continuous with respect to* x.

Proof. Using (10.4.11) we have

$$F_1(+0, t) - F_1(-0, t) = \int_0^t \Big(F_0(+0, t - \tau) - F_0(-0, t - \tau) \Big) V(0) \, d\tau. \tag{10.4.12}$$

Since the point $x = 0$ belongs to Δ, we see that $V(0) = 0$. Hence, $F_1(+0, t) - F_1(-0, t) = 0$. The theorem is proved. □

It follows from (7.1.2) that

$$\mu(z, t) = \mu(z, t - \tau) \mu(z, \tau). \tag{10.4.13}$$

Due to (10.4.13) and the convolution formula for Stieltjes–Fourier transform [9, Chap. 4], the relation

$$F_0(x, t) = \int_{-\infty}^{\infty} F_0(x - \xi, t - \tau) \, d_\xi F_0(\xi, \tau) \tag{10.4.14}$$

is valid. Using (10.4.11) and (10.4.14) we have

$$0 \leq F_n(x,t) \leq \frac{t^n F_0(x,t)}{n!}. \tag{10.4.15}$$

Hence, the series

$$F(x,t,u) = \sum_{n-0}^{\infty} (-1)^n u^n F_n(x,t) \tag{10.4.16}$$

converges. The probabilistic meaning of $F(x,t,u)$ is shown by the following relation [31, Chap. 4]:

$$E\left\{ \exp\left(-u \int_0^t V(X_\tau)\, d\tau \right), c_1 < X_t < c_2 \right\} = F(c_2,t,u) - F(c_1,t,u). \tag{10.4.17}$$

The inequality $V(x) \geq 0$ and relation (10.4.17) imply that the function $F(x,t,u)$ monotonically decreases with respect to the variable u and monotonically increases with respect to the variable x. Hence, the relations

$$0 \leq F(x,t,u) \leq F(x,t,0) = F_0(x,t) \tag{10.4.18}$$

are valid. In view of (10.4.18) the Laplace transform

$$\Psi(x,s,u) = \int_0^\infty e^{-st} F(x,t,u)\, dt, \quad s > 0 \tag{10.4.19}$$

is correct. Since the function $F(x,t,u)$ monotonically decreases with respect to u, this is also valid for the function $\Psi(x,s,u)$. Hence, the limits

$$F_\infty(x,t) = \lim F(x,t,u), \quad \Psi_\infty(x,s) = \lim \Psi(x,s,u), \quad u \to \infty, \tag{10.4.20}$$

exist. It follows from (10.4.17) that

$$p(t,\Delta) = P(X_\tau \in \Delta, 0 < \tau < t) = \int_\Delta d_x F_\infty(x,t). \tag{10.4.21}$$

Thus, we have

$$\int_0^\infty e^{-st} p(t,\Delta)\, dt = \int_\Delta d_x \Psi_\infty(x,s). \tag{10.4.22}$$

2. Relations (10.4.22) imply the next statement.

Proposition 10.4.5. *Let Conditions* 10.4.1 *be fulfilled. Then the function* $\Psi_\infty(x, s)$ *is monotonically increasing with respect to* x *and continuous at* $x \neq 0$ *for each* $s > 0$.

Using (10.4.10) and (10.4.19) we derive the following assertion.

Proposition 10.4.6. *Let Conditions* 10.4.1 *be fulfilled. Then*

$$\Psi(+0, s, u) - \Psi(-0, s, u) = \Psi_\infty(+0, s) - \Psi_\infty(-0, s) = \frac{1}{s + \Omega}, \quad s > 0. \quad (10.4.23)$$

The behavior of $\Psi_\infty(x, s)$ when $s = 0$ we shall consider (separately from the case $s > 0$) using the following Hengartner and Theodorescu result [23].

Theorem 10.4.7. *Let* X_t *be a Lévy process. Then for any finite interval* K *the estimate*

$$P\big(X_t \in K\big) = O\left(t^{-1/2}\right) \quad as \quad t \to \infty \qquad (10.4.24)$$

is valid.

Hence, we have the following assertion [98].

Theorem 10.4.8. *Let* X_t *be a Lévy process. Then for any integer* $n > 0$ *the estimate*

$$p(t, \Delta) = O(t^{-n/2}) \quad as \quad t \to \infty \qquad (10.4.25)$$

is valid.

We need the following particular case of (10.4.22):

$$\int_0^\infty p(t, \Delta)\, \mathrm{d}t = \int_\Delta \mathrm{d}_x \Psi_\infty(x, 0). \qquad (10.4.26)$$

According to (10.4.25) the integral on the left-hand side of (10.4.26) converges.

3. *Let us study the functions* $F(x, t, u)$ *and* $\Psi(x, s, u)$ *in detail.* $\impliedby$
According to (10.4.11) and (10.4.16) the function $F(x, t, u)$ satisfies the equation

$$F(x, t, u) + u \int_0^t \int_{-\infty}^\infty F_0(x - \xi, t - \tau) V(\xi)\, \mathrm{d}_\xi F(\xi, \tau, u)\, \mathrm{d}\tau = F_0(x, t). \quad (10.4.27)$$

Applying the Laplace transform to both parts of (10.4.27), and using (10.4.19) and the convolution property [9, Chap. 4], we obtain

$$\Psi(x, s, u) + u \int_{-\infty}^\infty \Psi_0(x - \xi, s) V(\xi)\, \mathrm{d}_\xi \Psi(\xi, s, u) = \Psi_0(x, s), \qquad (10.4.28)$$

where

$$\Psi_0(x, s) = \int_0^\infty e^{-st} F_0(x, t) \, dt. \tag{10.4.29}$$

It follows from (10.1.1) and (10.4.29)that

$$\int_{-\infty}^\infty e^{ixp} \, d_x \Psi_0(x, s) = \frac{1}{s + \lambda(p)}. \tag{10.4.30}$$

According to (10.4.28) and (10.4.29) we have

$$\int_{-\infty}^\infty e^{ixp} \big(s + \lambda(p) + uV(x)\big) \, d_x \Psi(x, s, u) = 1. \tag{10.4.31}$$

Now we introduce the function

$$h(p) = \frac{1}{2\pi} \int_\Delta e^{-ixp} f(x) \, dx, \tag{10.4.32}$$

where the function $f(x)$ belongs to C_Δ. Here C_Δ stands for the set of functions $g(x)$ on $L^2(\Delta)$ such that

$$g(a_k) = g(b_k) = g'(a_k) = g'(b_k) = 0, \quad 1 \le k \le n, \tag{10.4.33}$$

and $g''(x)$ is continuous function.

Multiplying both parts of (10.4.31) by $h(p)$ and integrating them with respect to p $(-\infty < p < \infty)$ we deduce the equality

$$\int_{-\infty}^\infty \int_{-\infty}^\infty e^{ixp} \big(s + \lambda(p)\big) h(p) \, d_x \Psi(x, s, u) \, dp = f(0). \tag{10.4.34}$$

We have used the relations

$$V(x) f(x) = 0, \quad -\infty < x < \infty, \tag{10.4.35}$$

$$\frac{1}{2\pi} \lim \int_{-N}^N \int_\Delta e^{-ixp} f(x) \, dx \, dp = f(0), \quad N \to \infty. \tag{10.4.36}$$

Since the function $F(x, t, u)$ monotonically decreases with respect to u, the function $\Psi(x, s, u)$ (see (10.4.19)) monotonically decreases with respect to u as well. Hence, there exist the limits

$$F_\infty(x, s) = \lim F(x, s, u), \quad \Psi_\infty(x, s) = \lim \Psi(x, s, u), \quad u \to \infty. \tag{10.4.37}$$

Using relations (10.1.2) and (10.1.6) we deduce that

$$\lambda(z) \int_{-\infty}^{\infty} e^{-iz\xi} f(\xi) \, d\xi = - \int_{-\infty}^{\infty} e^{-iz\xi} \big(Lf(\xi) \big) \, d\xi. \tag{10.4.38}$$

Relations (10.4.34), (10.4.37) and (10.4.38) imply the following assertion.

Theorem 10.4.9. *Let Conditions 10.4.1 be fulfilled. Then the relation*

$$\int_{\Delta} \big(sI - L_\Delta \big) f \, d_x \Psi_\infty(x, s) = f(0) \tag{10.4.39}$$

is valid.

Remark 10.4.10. For the Lévy processes of class II equality (10.4.39) was deduced in the Chapter 8.

4. Now, we shall find the relation between $\Psi_\infty(x, s)$ and $\Phi(0, x)$.

Theorem 10.4.11. *Let Conditions 10.4.1 and inequality (10.2.6) be fulfilled. Then in the space D_Δ^* there is one and only one function*

$$\Psi(x, s) = \big(I + sB^* \big)^{-1} \Phi(0, x), \tag{10.4.40}$$

which satisfies relation (10.4.39).

Proof. In view of (10.2.2) we have

$$-BL_\Delta f = f, \quad f \in C_\Delta. \tag{10.4.41}$$

Relations (10.4.40) and (10.4.41) imply that

$$\Big\langle \big(sI - L_\Delta \big) f, \psi(x, s) \Big\rangle_\Delta = -\big\langle (I + sB) L_\Delta f, \psi \big\rangle_\Delta = -\big\langle L_\Delta f, \Phi(0, x) \big\rangle_\Delta. \tag{10.4.42}$$

Since $\Phi(0, x) = B^* \sigma(x)$, then according to (10.4.40) and (10.4.42) relation (10.4.39) is valid.

Let us suppose that in D_Δ^* there is another function $\Psi_1(x, s)$ satisfying (10.4.39). Then the equality

$$\Big\langle \big(sI - L_\Delta \big) f, \varphi(x, s) \Big\rangle_\Delta = 0, \quad \varphi = \Psi - \Psi_1, \tag{10.4.43}$$

is valid. We write relation (10.4.43) in the form

$$\Big\langle L_\Delta f, \big(I + sB^* \big) \varphi \Big\rangle_\Delta = 0. \tag{10.4.44}$$

The range of L_Δ is dense in D_Δ. Hence, in view of (10.4.44) we have $\varphi = 0$. The theorem is proved. $\qquad\square$

It follows from relations (10.4.39) and (10.4.40) that

$$\Psi_\infty(x, s) = \big(I + sB^* \big)^{-1} \Phi(0, x). \tag{10.4.45}$$

10.5 Long time behavior

In this section we study the asymptotic behavior of $p(t, \Delta)$, when $t \to \infty$ and $X_t \in I_c$. Recall that the asymptotic behavior of $p(t, \Delta)$, when a Lévy process X_t belongs to the class II, was investigated in Section 8.5. In order to study $p(t, \Delta)$, we apply the Kreĭn–Rutman theorem (see Theorem 8.5.3). The corresponding operator T_1 has the form (10.2.9), the cone K consists of non-negative continuous real functions $g(x) \in D_\Delta$ and the cone K^* consists of monotonically increasing bounded real functions $h(x) \in D_\Delta^*$.

The spectrum of the operator $B = (1/\Omega)(I + T_1)$ is situated in the domain $\operatorname{Re} z > 0$. The eigenvalue $\mu_1 = (1/\Omega)(1 + \lambda_1)$ of the operator B is larger in modulus than any other eigenvalue μ_k ($k > 1$) of B. Introduce the domain D_ε:

$$\left| z - \frac{1}{\Omega} \right| < \varepsilon, \quad 0 < \varepsilon < \frac{1}{\Omega}. \tag{10.5.1}$$

We denote the boundary of the domain D_ε by Γ_ε. If z belongs to D_ε, then the relations

$$\operatorname{Re} \frac{1}{z} > c_c > 0 \tag{10.5.2}$$

hold. Now, we formulate the main result of this section.

Theorem 10.5.1. *Let Lévy process X_t belong to the class I_c, let $0 \in \Delta$, and assume that the corresponding operator T_1 satisfies the following conditions:*

1. *Operator T_1 is compact in the Banach space D_Δ.*
2. *Operator T_1 has some point of spectrum different from zero.*
3. *Conditons 10.4.1 and inequality (10.2.6) are fulfilled.*
4. *ind $\lambda_1 = \operatorname{gmul} \lambda_1 = 1$.*

Then the asymptotic equality

$$p(t, \Delta) = e^{-t/\mu_1}(c_1 + o(1)), \quad t \to +\infty, \quad c_1 \geq 0 \tag{10.5.3}$$

holds. If $g_1(0) > 0$, then $c_1 > 0$.

Proof. Using (10.4.40) we obtain the equality

$$p(t, \Delta) = \frac{1}{2\pi} \int_{-\infty}^{\infty} \langle e^{iyt}, \Psi_\infty(x, iy) \rangle_\Delta \, dy, \quad t > 0. \tag{10.5.4}$$

Changing the variable $z = i/y$ we rewrite (10.5.4) in the form

$$p(t, \Delta) = \frac{1}{2i\pi} \int_{-i\infty}^{i\infty} \langle e^{-t/z}, (zI - B^*)^{-1} \Phi(0, x) \rangle_\Delta \frac{dz}{z}, \quad t > 0. \tag{10.5.5}$$

Since the operator T_1 is compact, only a finite number of the eigenvalues λ_k $(1 < k \leq m)$ of this operator does not belong to the domain D_ε. From formula (10.5.5) we deduce the relation

$$p(t, \Delta) = \sum_{k=1}^{m} \sum_{j=0}^{n_k-1} e^{-t/\lambda_k} t^j c_{k,j} + J, \qquad (10.5.6)$$

where n_k is the dimension of the largest Jordan block of T_1 associated to the eigenvalue λ_k (i.e., the index of λ_k),

$$J = -\frac{1}{2i\pi} \int_{\Gamma_\varepsilon} \frac{1}{z} e^{-t/z} \left\langle \mathbb{1}, (B^* - zI)^{-1} \Phi(0, x) \right\rangle_\Delta dz. \qquad (10.5.7)$$

The end of the proof coincides with the corresponding part of the proof of Theorem 8.5.5. $\qquad \square$

10.6   Example

1. As an example let us consider the case where

$$\nu'(x) = p^2 e^{-p|x|}, \quad p > 0, \quad \Delta = [0, \omega], \quad A = 0, \quad \gamma = 0. \qquad (10.6.1)$$

Using (10.1.17) we have

$$L_\Delta f = -2p \left(f(x) - \frac{p}{2} \int_0^\omega e^{-p|x-y|} f(y)\, dy \right). \qquad (10.6.2)$$

Since the condition (10.2.6) is fulfilled, we see that the operator L_Δ^{-1} has the form (10.2.7), where the operator T_1 is given by the relation

$$T_1 f = \int_0^\omega \gamma(x, t) f(t)\, dt. \qquad (10.6.3)$$

It follows from (10.2.7) and (10.6.2) that

$$\gamma(x, t) - \frac{p}{2} e^{-p|x-t|} - \frac{p}{2} \int_0^\omega e^{-p|x-y|} \gamma(y, t)\, dy = 0. \qquad (10.6.4)$$

According to (10.6.4) we have

$$\frac{\partial^2 \gamma}{\partial x^2} = 0, \quad x \neq t. \qquad (10.6.5)$$

From (10.6.5) we obtain that

$$\gamma(x,t) = c_1(t) + c_2(t)x, \quad x > t, \tag{10.6.6}$$
$$\gamma(x,t) = c_3(t) + c_4(t)x, \quad x < t. \tag{10.6.7}$$

Using (10.6.6), (10.6.7) and the equality $\gamma(x,t) = \gamma(t,x)$ we deduce that

$$\gamma(x,t) = (\alpha_1 + \alpha_2 t) + (\beta_1 + \beta_2 t)x, \quad t < x. \tag{10.6.8}$$

In view of (10.6.6) and relations

$$\gamma(x,0) - \frac{p}{2}e^{-px} - \frac{p}{2}\int_0^\omega e^{-p|x-y|}\gamma(y,0)\,dy = 0, \tag{10.6.9}$$

$$\left.\frac{\partial\gamma(x,t)}{\partial t}\right|_{t=0} - \frac{p^2}{2}e^{-px} - \frac{p}{2}\int_0^\omega e^{-p|x-y|}\left.\frac{\partial\gamma(y,t)}{\partial t}\right|_{t=0}dy = 0 \tag{10.6.10}$$

we obtain

$$\alpha_1 = \frac{p(1+p\omega)}{2+p\omega}, \quad \beta_1 = -\frac{p^2}{2+p\omega}, \quad \alpha_2 = p\alpha_1, \quad \beta_2 = p\beta_1. \tag{10.6.11}$$

Thus, the corresponding operator T_1 has the form (10.6.3), where the kernel $\gamma(x,t)$ is defined by the relations (10.6.8), (10.6.11) and the equality $\gamma(x,t) = \gamma(t,x)$.

2. Let us find the eigenvalues and eigenfunctions of the operator T_1 (case (10.6.1)). The eigenfunctions $f(x)$ and eigenvalues λ satisfy the relations

$$\int_0^\omega \gamma(x,t)f(t)\,dt = \lambda f(x), \quad \int_0^\omega \frac{\partial}{\partial x}\gamma(x,t)f(t)\,dt = \lambda f'(x). \tag{10.6.12}$$

It follows from (10.6.12) that

$$\lambda f''(x) = -p^2 f. \tag{10.6.13}$$

Hence,

$$f(x,\lambda) = c_1(\lambda)\sin\frac{xp}{\sqrt{\lambda}} + c_2(\lambda)\cos\frac{xp}{\sqrt{\lambda}}. \tag{10.6.14}$$

According to (10.6.12) and (10.6.14) we have

$$c_1(\lambda)a(\lambda) + c_2(\lambda)b(\lambda) = \lambda c_2(\lambda), \tag{10.6.15}$$
$$c_1(\lambda)a(\lambda) + c_2(\lambda)b(\lambda) = \sqrt{\lambda}c_1(\lambda), \tag{10.6.16}$$

where

$$a(\lambda) = \frac{1}{2+p\omega}\left(-\sqrt{\lambda}\cos\frac{p\omega}{\sqrt{\lambda}} - \lambda\sin\frac{p\omega}{\sqrt{\lambda}} + (1+p\omega)\sqrt{\lambda}\right), \tag{10.6.17}$$

$$b(\lambda) = \frac{1}{2+p\omega}\left(\sqrt{\lambda}\sin\frac{p\omega}{\sqrt{\lambda}} - \lambda\cos\frac{p\omega}{\sqrt{\lambda}} + \lambda\right). \tag{10.6.18}$$

In view of (10.2.7) the eigenvalues of the operator B are defined by the relations (10.6.15) and (10.6.16), that is, by

$$\sqrt{\lambda}a(\lambda) + b(\lambda) = \lambda. \tag{10.6.19}$$

Equalities (10.6.17)–(10.6.19) imply that

$$\tan\frac{p\omega}{\sqrt{\lambda}} = \frac{2\sqrt{\lambda}}{1-\lambda}. \tag{10.6.20}$$

The following table gives the maximal positive roots of the equation (10.6.20).

$$\begin{pmatrix} p\omega & \pi/4 & \pi/3 & \pi/2 & 2\pi/3 & \pi \\ \lambda & 0.445 & 0.617 & 0.162 & 1.433 & 2.454 \end{pmatrix} \tag{10.6.21}$$

The author is grateful to I. Tydniouk for his calculations presented in (10.6.21).

10.7   Discrete Lévy measure

Let us consider the case where the Lévy measure $\nu(u)$ is discrete. We denote the gap points by ν_k ($1 \le k \le n$) and the corresponding gaps by $\sigma_k > 0$. Then formula (10.1.17) takes the form

$$L_\Delta f = -\Omega f(x) + \sum_{k=1}^{n} f(\nu_k + x)\sigma_k, \quad \Omega \ge \sum_{k=1}^{n}\sigma_k, \tag{10.7.1}$$

and the operator T is given by the relation

$$Tf = \sum_{k=1}^{n} f(\nu_k + x)\sigma_k, \quad x \in \Delta. \tag{10.7.2}$$

We note that $f(x) = 0$ if $x \notin \Delta$. Relation (10.7.2) implies the following assertion.

Proposition 10.7.1. *The operator T has the following properties*:

1. *The operator T maps bounded functions into bounded functions.*

2. *The operator T maps non-negative functions into non-negative functions.*

Remark 10.7.2. The methods of Section 10.6 can be applied to the discrete case. In this case, Lebesgue–Stieltjes integrals should be used instead of Stieltjes integrals.

Chapter 11

Open Problems

11.1 Triangular form

1. We consider bounded linear operators acting in a Hilbert space H. First, we formulate some definitions and results (without proofs) from our papers [71] and [72].

Definition 11.1.1. The maximal invariant subspace of the operators A and A^*, on which the equality $AA^* = A^*A$ is fulfilled, is called the additional component.

We note that M.S. Livšic [52] introduced the additional component using the equality $A = A^*$.

Definition 11.1.2. We say that the operator A belongs to the class $\mathcal{T}$ if for any two invariant subspaces H_1 and H_2 ($H_1 \subset H_2$, $\dim(H_2 \ominus H_1) > 1$) of the operator A there exists a third invariant subspace H_3 of A such that $H_1 \subset H_3 \subset H_2$ and $H_1 \neq H_3$, $H_3 \neq H_2$.

We also consider bounded triangular operators $\vec{A}$ of the form

$$\vec{A}f = \frac{\mathrm{d}}{\mathrm{d}x} \int_0^x K(x,t)f(t)\,\mathrm{d}t, \quad 0 \leq x \leq 1, \tag{11.1.1}$$

acting in the spaces $L_m^2(0,1)$ ($m \leq \infty$).

Theorem 11.1.3. *For every operator A from the class $\mathcal{T}$ there exists a triangular operator $\vec{A}$ of the form* (11.1.1) *and a unitary operator U with the following properties.*

1. *The operator U is a one-to-one mapping from the space $H \ominus C_A$ onto the space $L_m^2(0,1) \ominus C_{\vec{A}}$, where C_A and $C_{\vec{A}}$ are the additional components of A and $C_{\vec{A}}$, respectively.*

2. *The reduction of A onto $H \ominus C_A$ is transformed by U into the reduction of $\vec{A}$ onto $L_m^2(0,1) \ominus C_{\vec{A}}$, that is, we have*

$$\vec{A} = UAU^{-1} \quad \text{for} \quad f \in (L_m^2(0,1) \ominus C_{\vec{A}}). \tag{11.1.2}$$

Remark 11.1.4. Using analytic methods, M.S. Livšic proved in [71] that the operators of the class iΩ can be reduced to the triangular form. Recall that an operator A belongs to the class iΩ if

$$\sum_j |\lambda_j| < \infty, \qquad (11.1.3)$$

where λ_j are the eigenvalues of the operator $(A - A^*)/i$.

Remark 11.1.5. The formulated above Theorem 11.1.3 is an essential generalization of the corresponding Livšic's result. We derived it using geometric methods.

It follows from the Aronszajn–Smith theorem [3] that all compact operators belong to the class $\mathcal{T}$.

Corollary 11.1.6. *Any compact operator A is, up to the additional component, unitary equivalent to a triangular operator $\vec{A}$ of the form* (11.1.1).

2. Let us consider operators A of the form

$$A = A_1 + iB, \qquad (11.1.4)$$

where A_1 is a bounded and self-adjoint operator and B is a self-adjoint operator from the Hilbert–Schmidt class.

Definition 11.1.7. We say that the operators A of the form (11.1.4) belong to the class $\mathcal{T}_1$.

Theorem 11.1.8. *Every operator belonging to the class $\mathcal{T}_1$ has a proper invariant subspace.*

Corollary 11.1.9. *If an operator A belongs to the class $\mathcal{T}_1$ then A also belongs to the class $\mathcal{T}$.*

Theorem 11.1.10. *Let the operator A belong to the class $\mathcal{T}_1$ and have purely real spectrum. Then A is, up to the additional component, unitary equivalent to a triangular operator of the form*

$$\vec{A}f = H(x)f(x) + \int\limits_0^x N(x,t)f(t)\,\mathrm{d}t, \quad 0 \le x \le 1. \qquad (11.1.5)$$

Here $H(x)$ and $N(x,t)$ are $m{\times}m$ matrix functions and

$$H(x) = H^*(x), \quad \sum_{i,j=1}^m \int\limits_0^1 \int\limits_0^1 |n_{i,j}(x,t)|^2 \,\mathrm{d}x\,\mathrm{d}t < \infty, \qquad (11.1.6)$$

where $n_{i,j}(x,t)$ are the entries of the matrix $N(x,t)$.

Theorem 11.1.11. *Let the operator A belong to the class $\mathcal{T}_1$ and let the spectrum of A be concentrated at zero. Then A is, up to the additional component, unitary equivalent to a triangular operator of the form*

$$\vec{A}f = \int_0^x N(x,t)f(t)\,dt, \quad 0 \le x \le 1, \tag{11.1.7}$$

where

$$\sum_{i,j=1}^m \int_0^1 \int_0^1 |n_{i,j}(x,t)|^2 \, dx\,dt < \infty. \tag{11.1.8}$$

The following important problem is very well known.

Problem 11.1.12. *Consider bounded linear operators A in a separable Hilbert space. Has every such A a non-trivial closed invariant subspace?*

If the answer is "yes", then every such operator A belongs to the class $\mathcal{T}$, that is, it can be reduced to a triangular form.

The next problem is a special case of Problem 11.1.12.

Problem 11.1.13. *Now, we consider operators A of the form*

$$A = A_1 + iB, \tag{11.1.9}$$

where A_1 is a bounded and self-adjoint operator and B is a self-adjoint and compact operator in a Hilbert space. Has every such A a non-trivial closed invariant subspace?

Again, if the answer is "yes", then every such operator A belongs to the class $\mathcal{T}$, that is, it can be reduced to a triangular form.

11.2 Quasi-potential operator B

Problem 11.2.1. *Construct explicitly quasi-potential operators for the stable processes such that*

$$\alpha = 1, \quad -1 \le \beta \le 1, \quad \beta \ne 0. \tag{11.2.1}$$

We note that quasi-potential operators B were constructed explicitly for all the cases besides (11.2.1) (see Section 8.7).

It is easy to prove that the quasi-potential operator B is sectorial in the space of continuous functions (see Section 8.4). However, the following problem is still open.

Problem 11.2.2. *Find conditions under which the quasi-potential operator B is strongly sectorial in the space of continuous functions.*

11.3   Analogue of the Weyl problem from spectral theory

We consider the case when a Lévy process X_t is stable and

$$0 < \alpha < 2, \quad \beta = 0, \quad \Delta = [-a, a]. \tag{11.3.1}$$

By λ_n we denote the eigenvalues of the corresponding quasi-potential operator B. An explicit expression for such an operator B is given in Section 8.7. Let us write the asymptotic equalities

$$\mu_n = (\pi n/2a)^\alpha \left(1 + o(1)\right), \quad n \to \infty, \quad \mu_n = 1/\lambda_n. \tag{11.3.2}$$

For the case $\alpha = 1$ asymptotics (11.3.2) was obtained in [36, 92] and for the case $\alpha \neq 1$ it was derived in [87, 92]. It follows from (11.3.2) that

$$N(\mu) = ((\mathrm{mes}\,\Delta)/\pi)\,\mu^{1/\alpha}(1 + o(1)), \quad \mu \to \infty, \tag{11.3.3}$$

where $N(\mu)$ is the number of the eigenvalues μ_n less or equal to μ.

The asymptotics of $N(\mu)$ is of interest in the case when Δ is a set of several segments $[a_k, b_k]$ (general type Δ). We note that, similar to the Weyl formula, the first term of the asymptotic formula (11.3.3) depends on the measure of Δ. It is possible that the second term of asymptotics in (11.3.3) depends on the measure of the boundary of Δ, that is, on the measure of the points a_k and b_k. Recall that the measure of a point is equal to 1 (see [85, Section 4]).

Hypothesis A. *The second term of the asymptotics* (11.3.3) *depends on the number of different segments in the domain* Δ.

Hypothesis B. *An analogue of the asymptotic Weyl formula is valid, that is,*

$$N(\mu) = ((\mathrm{mes}\,\Delta)/\pi)\,\mu^{1/\alpha} - (q/2) + o(1), \quad \mu \to \infty, \tag{11.3.4}$$

where q *is the number of different segments in the domain* Δ.

Problem 11.3.1. *Find the second term of the asymptotics* (11.3.3). *Are Hypotheses A and* B *true?*

11.4   Fractional integrals of purely imaginary order

Consider a fractional integral operator $\mathcal{J}^{i\alpha}$, $\alpha = \overline{\alpha}$ given by (3.4.1):

$$\mathcal{J}^{i\alpha} f = \frac{1}{\Gamma(i\alpha + 1)} \frac{\mathrm{d}}{\mathrm{d}x} \int_0^x (x - t)^{i\alpha} f(t)\,\mathrm{d}t, \quad 0 \le x \le \omega.$$

The spectrum of this operator is described in Remark 3.4.1 (see also [81]). The operator $\mathcal{J}^{i\alpha}$, $\alpha = \overline{\alpha}$, is not unicellular [81]. We studied the properties of this interesting and important operator in the paper [81]. However, the following problem is open.

Problem 11.4.1. *Describe all the invariant subspaces of the operator* $\mathcal{J}^{i\alpha}$, $\alpha = \overline{\alpha}$.

Commentaries and Remarks

1. Formula (1.1.10) is a special case of the operator identity of the form

$$AS - SA = \Pi_1 \Pi_2^*, \tag{1}$$

which is a generalization of the well-known notion of a node (see M.S. Livšic [53] and then M.S. Brodskii [10]). The identities of the form (1) are used for solving problems in numerous domains (system theory [90], factorization problems [90], interpolation theory [28, 93], the inverse spectral problem [90, 94] and theory of nonlinear integrable equations [91, 94]). There are close ties between all these problems and between corresponding results.

2. Apart from the regularization method, there was in fact no approach to the study of equations of the first kind. The operator identities method permits us to construct a theory of equations of the form

$$Sf = \varphi \tag{2}$$

without a requirement of invertibility of the operator S [83, 84, 86].

This approach is presented in all details in Chapter 2 for operators S with a difference kernel and in Chapter 5 for operators with a W-difference kernel.

3. The method of inverting operators with a difference kernel considered in this book was first introduced in connection with the problem of reducing Volterra operators to their simplest form [76]. In this book the inversion method was set forth independently of Volterra operators. This permitted us to omit the requirement on S to be positive.

4. As it was shown in Chapter 1 and Chapter 2, a bounded in $L^p(0, \omega)$ $(p \geq 1)$ operator S with a difference kernel admits a representation

$$Sf = \frac{\mathrm{d}}{\mathrm{d}x} \int_0^\omega s(x - t) f(t) \, \mathrm{d}t, \quad f \in L^p(0, \omega). \tag{3}$$

The representation of the operator S in the form (3) allows us to consider uniformly different classes of the operators S and of the corresponding operator equations

$$Sf = \varphi. \tag{4}$$

In order to demonstrate this, we give a list of the main special cases of formula (3) and also indicate the chapters, where these special cases are considered:

$$Sf = \sum_{j=N}^{M} \mu_j f(x - x_j) + \int_0^\omega K(x - t) f(t)\, \mathrm{d}t, \tag{5}$$

where $f(x) = 0$ $x \overline{\in} [0, \omega]$, $K(x) \in L(-\omega, \omega)$ (Chapter 1);

$$Sf = \mu f(x) + \int_0^\omega K(x - t) f(t)\, \mathrm{d}t, \tag{6}$$

where $K(x) \in L(-\omega, \omega)$, $\mu \neq 0$ (Chapters 1–3, 9);

$$Sf = \int_0^\omega K(x - t) f(t)\, \mathrm{d}t, \tag{7}$$

where $K(x) \in L(-\omega, \omega)$ (Chapters 1–4, 8, 9);

$$Sf = \int_0^\omega \left(\frac{1}{x - t} + K(x - t) \right) f(t)\, \mathrm{d}t, \tag{8}$$

where $K(x) \in L(-\omega, \omega)$ (Chapter 3);

$$Sf = \frac{1}{\Gamma(i\alpha + 1)} \frac{\mathrm{d}}{\mathrm{d}x} \int_0^x (x - t)^{i\alpha} f(t)\, \mathrm{d}t \tag{9}$$

(Chapters 1, 3).

5. **Explicit solutions of equations.** Explicit solutions of the equation

$$Sf = \int_0^\omega k(x - t) f(t)\, \mathrm{d}t = \varphi(x) \tag{10}$$

are constructed in the cases

$$k(x) = \frac{1 - \beta \operatorname{sgn}(x)}{|x|^{\alpha - 1}}, \quad -1 \le \beta \le 1, \quad 0 < \alpha < 2 \quad \text{(Chapters 3 and 9),} \tag{11}$$

$$k(x) = \ln \frac{b}{2|x|}, \qquad b > 0 \quad \text{(Chapters 3 and 9),} \tag{12}$$

$$k(x) = \ln \frac{\sin(b/2)}{2 \sin(|x|/2)}, \qquad b > 0 \quad \text{(Chapter 9),} \tag{13}$$

$$k(x) = \ln \frac{\sinh(b/2)}{2 \sinh(|x|/2)}, \qquad b > 0 \quad \text{(Chapter 9).} \tag{14}$$

6. Convolution form, Chapter 7. Convolution form representation of the generator L was obtained, first, for stable processes [92]. Later we proved that the convolution form representation of the generator L is valid for a broad class of Lévy processes [97]. In Chapter 7 of this book we prove a general fact:

If X_t is a Lévy process, then the corresponding generator L can be represented in the convolution form.

7. Quasi-potential operator B, Chapter 8. M. Kac and H. Pollard [33] showed that the study of stable Lévy processes is connected with the solution of integro-differential equations (see (8.3.26)). M. Kac solved such equations for the case of Cauchy symmetric processes [30]. Later H. Widom solved such equations for all symmetric stable Lévy processes [111]. We developed the results of M. Kac and H. Widom further and proved their analogs for a wide class of Lévy processes [97]. Here we construct the mentioned above integro-differential equations for all Lévy processes. The solution of these equations is given in terms of the quasi-potential operator B (which is obtained by inversion of the corresponding integro-differential operator). In Chapter 8 we prove the following general fact:

Let X_t be a Lévy process. Then the corresponding quasi-potential operator B is a bounded operator in the Banach space of continuous functions.

8. Wiener processes, Chapter 8. The Wiener process is an important special case of the stable processes (namely, the case where $\alpha = 2$). Our approach was applied to this case in the book [97].

9. Proposition 8.3.7, Chapter 8. Consider the stochastic process

$$Y_t = \int_0^t V(X_\tau)\, d\tau. \tag{15}$$

Denote by $R_t(y)$ the distribution function of Y_t. Then we have

$$E\left[e^{-uY_t}\right] = \int_0^t e^{-uy}\, d_y R_t(y), \tag{16}$$

where $R_t(+0) - R_t(0) = p(t, \Delta)$. It follows from (16) that

$$\lim_{u \to \infty} E\left[e^{-uY_t}\right] = p(t, \Delta). \tag{17}$$

Bibliography

[1] Alexandrov V.M., Mkhitaryan, S.M., *Contact Problems for Bodies with Thin Covers and Layers* (Russian), Nauka, Moscow, 1983.

[2] Ambartsumian V.A., *Scientific Works* (Russian), vol. 1, Acad. Sci. ArmSSR, Erevan, 1960.

[3] Aronszajn N., Smith K.T., *Invariant Subspaces of Completely Continuous Operators,* Annals of Mathematics, 60, 2, 345–350, 1954.

[4] Arutunyan N.K., *Plane Contact Problem of the Theory of the Creep* (Russian), Appl. Math. and Mech., 23, 5, 901–924 , 1959.

[5] Bart H., Gohberg I., Kaashoek M.A., *Convolution Equations and Linear Systems*, Integral Equations Operator Theory, 5, 283–340, 1982.

[6] Bart H., Gohberg I., Kaashoek M.A., *The Coupling Method for Solving Integral Equations*, pp. 39–73, in: Operator Theory Adv. Appl., vol. 12, Birkhäuser, Basel, 1984.

[7] Bateman H., Erdelyi, A., *Higher Transcendental Functions*, Mc Graw-Hill Book Company, New York – London, 1953.

[8] Bertoin J., *Lévy Processes*, Cambridge Univ. Press, Cambridge, 1996.

[9] Bochner S., *Vorlesungen über Fouriersche Integrale*, Chelsea Publishing Company, New York, 1948.

[10] Brodskii M.S., *Triangular and Jordan Representations of Linear Operators*, Transl. Math. Monographs, vol. 32, Amer. Math. Soc., Providence, RI, 1971.

[11] Buslaev A.G., *Solution of the Contact Interaction Equation* (Russian), Dokl. Akad. Nauk Ukr. SSR, 7, 3–8, 1989.

[12] Carleman T., *Zur Theorie der linearen Integralgleichungen*, Math. Z., 9, 196–217, 1921.

[13] Chandrasekhar S., *Radiative Transfer*, Oxford University Press, London, 1950.

[14] Fock V.A., *On Certain Integral Equations of Mathematical Physics* (Russian), Mat. Sbornik, 14, 3–50 , 1944.

[15] Franks L.E., *Signal Theory*, Prentice-Hall: Englewood Cliffs, NJ, 1969.

[16] Fritzsche B., Kirstein B., Roitberg I.Ya., Sakhnovich A.L., *Recovery of Dirac System from the Rectangular Weyl Matrix Function*, Inverse Problems, 28, 015010, 18 p., 2012.

[17] Gakhov F.D., *Boundary-value Problems*, Pergamon Press, Oxford, 1966.

[18] Gelfand I.M., Shilov G.E., *Generalized Functions, I*, Academic Press, New York–London, 1964.

[19] Gelfand I.M., Vilenkin N.Y., *Some Applications of Harmonic Analysis* (Russian), Fizmatgiz, Moscow, 1961.

[20] Gladwell G.M., *Contact Problems in the Classical Theory of Elasticity*, Sijthoff & Noordhoff Intern. Publ., Alphen aan den Rijn, The Netherlands, 1980.

[21] Gohberg I., Kreĭn M.G., *Theory and Applications of Volterra Operators in Hilbert Space*, Transl. Math. Monographs, vol. 24, Amer. Math. Soc., Providence, RI, 1970.

[22] Gohberg I., Sementsul A.A., *On the Inversion of Finite Toeplitz Matrices and their Continuous Analogues* (Russian), Mat. Issled., 7, 2, 201–223 , 1972.

[23] Hengartner W., Theodorescu R., *Concentration Functions*, Academic Press, New York, 1973.

[24] Hönl H., Maue A.W., Westpfahl D.D.K., *Theorie der Beugung*, in: Kristalloptik, Beugung/Crystal Optics, Diffraction, pp. 218–573, Springer, Berlin-Heidelberg, 1961.

[25] Hille E., Phillips R., *Functional Analysis and Semigroups*, Amer. Math. Soc., Providence, RI, 1957.

[26] Hopf E., *Mathematical Problems of Radiative Equilibrium*, Cambridge Tracts in Math. and Math. Phys., 31, Cambridge Univ. Press, London, 1934.

[27] Ivanchenko T.S., Sakhnovich L.A., *An Operator Approach to the Investigation of Interpolation Problems* (Russian), Dep. at Ukr. NIINTI, No. 701, 63 p., Odessa, 1985.

[28] Ivanchenko T.S., Sakhnovich L.A., *An Operator Approach to V.P. Potapov's Scheme for the Investigation of Interpolation Problems* (Russian), Ukrain. Mat. Zh., 39, 5, 573–578, 1987. Translated in Ukrainian Math. J., 39, 5, 464–469, 1987.

[29] Ivanov V.V., *Radiation Transfer and the Spectra of Celestial Bodies* (Russian), Nauka, Moscow, 1969.

[30] Kac M., *On Some Connections Between Probability Theory and Differential and Integral Equations*, in: Proc. Sec. Berkeley Symp. Math. Stat. and Prob., pp. 189–215, Berkeley, 1951.

[31] Kac M., *Probability and Related Topics in Physical Sciences*, Lectures in Applied Mathematics, vol. 1a, Amer. Math. Soc., Providence, RI, 1957.

[32] Kac M., *Some Stochastic Problems in Physics and Mathematics*, Collected Lectures in Pure and Applied Science, no. 2. (Hectographed), Magnolia Petroleum Company, Dallas, 1957.

[33] Kac M., Pollard, H., *On the Distribution of the Maximum of Partial Sums of Independent Random Variables*, Canad. J. Math., 2, 375–384, 1950.

[34] Keldysh M.V., Lavrent'ev M.A., *On the Motion of a Wing under the Surface of a Heavy Liquid* (Russian), in: Proc. Conf. Resistance of Waves, pp. 31–64, ZAGI, Moscow, 1937.

[35] Kober H., *On a Theorem of Schur and of Fractional Integrals of Purely Imaginary Order*, Trans. Amer. Math. Soc., 50, 1, 160–174, 1941.

[36] Kogan K.M., Sakhnovich L.A., *The Spectral Asymptotic of a Certain Singular Integro-differential Operator* (Russian), Differ. Uravn., 20, 8, 1444–1447, 1984.

[37] Kostyukov A.A., *Theory of Ship Waves and Wave Resistance* (Russian), Sudpromgiz, Leningrad, 1959.

[38] Kreĭn M.G., *On a New Method of Solving Linear Integral Equations of the First and Second Kind* (Russian), Dokl. Akad. Nauk SSSR, 100, 413–416, 1955.

[39] Kreĭn M.G., *Continuous Analogues of Propositions on Polynomials Orthogonal on the Unit Circle* (Russian), Dokl. Akad. Nauk SSSR, 105, 637–640, 1955.

[40] Kreĭn M.G., *Integral Equations on a Halfline with a Kernel Depending on the Difference of the Arguments* (Russian), Uspekhi Mat. Nauk, 13,5, 3–120, 1958.

[41] Kreĭn M.G., *Introduction to the Geometry of Indefinite J-Spaces and to the Theory of Operators in Those Spaces* (Russian), in: Proc. of Second Math. Summer School, I, pp. 15–92, Naukova Dumka, Kiev, 1965.

[42] Kreĭn M.G., Langer H., *Some Propositions on Analytic Matrix Functions Related to the Theory of Operators in the Space Π_1*, Acta Sci. Math. (Szeged), 43, 181–205, 1981.

[43] Kreĭn M.G., Nudelman P.Y., *Approximation of Functions in $L^2(\omega_1, \omega_2)$ by the Transfer Functions of Linear Systems with Minimal Energy* (Russian), Problems of Information Transmission, 11, 2, 124–142, 1975.

[44] Kreĭn M.G., Naimark M.A., *The Method of Symmetric and Hermitian Forms in the Theory of Separation of Roots of Algebraic Equations* (Russian), ONTI, Kharkov, 1936. Translated in Linear and Multilinear Algebra, 10, 4, 265–308, 1981.

[45] Kreĭn M.G., Rutman M.A., *Linear Operators Leaving Invariant a Cone in a Banach Space* (Russian), Uspehi Matem. Nauk (N.S.) 3, 1(23), 3–95, 1948. Translated in Amer. Math. Soc. Transl., vol. 26, Amer. Math. Soc., New York, 1950.

[46] Landau H.J., Pollak H.O., *Prolate Spheroidal Wave Functions, Fourier Analysis and Uncertainty, II*, Bell System Tech. J., 40, 65–84, 1961.

[47] Landau H.J., Pollak H.O., *Prolate Spheroidal Wave Functions, Fourier Analysis and Uncertainty, III, The dimension of the space of essentially time- and band-limited signals*, Bell System Tech. J., 41, 1295–1336, 1962.

[48] Larson D.R., *Nest Algebras and Similarity Transformations*, Annals of Mathematics, 121, 409–427, 1985.

[49] Levin B.R., *Theoretical Foundations of Statistical Radio Engineering* (Russian), Soviet Radio, Moscow, 1968.

[50] Levitan B.M., *Some Questions of the Theory of Almost Periodic Functions*, Amer. Math. Soc. Transl., vol. 28, Amer. Math. Soc., 1950.

[51] Levitan B.M., *Almost Periodical Functions* (Russian), Nauka, Moscow, 1953.

[52] Livšic M.S., *On Spectral Decomposition of Linear Non-Selfadjoint Operators*, Mat. Sb., N. Ser., 34(76), 4, 145–199 (Russian), 1954. Translated in Amer. Math. Soc. Transl. (2), 5, 67–114, 1957.

[53] Livšic M.S., *Operators, Oscillations, Waves (Open Systems)*, Transl. Math. Monographs, vol. 34, Amer. Math. Soc., Providence, 1973.

[54] Lumer G, Philips R.S., *Dissipative Operators in Banach Space*, Pacific J. Math., 11, 679–698, 1961.

[55] Muskhelishvili N.I., *Singular Integral Equations*, P. Noordhoff Ltd., Groningen, 1961.

[56] Olshevsky V., Sakhnovich L.A., *Optimal Prediction Problems for Generalized Stationary Processes*, in: Operator Theory Adv. Appl., vol. 160, pp. 257–266, Birkhäuser, Basel, 2005.

[57] Olshevsky V., Sakhnovich L.A., *Matched Filtering for Generalized Stationary Processes*, IEEE Trans. Inform. Theory, 51, 9, 3308–3313, 2005.

[58] Polishchuk I.M., Lebedev Yu.N., *Some Properties of the Problem of Diffraction on a Strip*, Radio Engrg. Electron Phys., 22, 12, 2489–2493, 1977.

[59] Pozin S.M., *On the Structure of Solutions of Integral Equations with Difference Kernels* (Russian), Dissertation, Odessa, 1984.

[60] Pozin S.M., Sakhnovich L.A., *Behavior of Solutions of an Equation with Difference Kernel* (Russian), Izv. Akad. Nauk Armyan. SSR Ser. Mat., 17, 5, 376–386, 1982.

[61] Radon J., *Über lineare Funktionaltransformationen und Funktionalgleichungen*, Wien. Anz. 56, 189, 1919; Ber. Akad. Wiss. Wien, 128, 1083–1121, 1919.

[62] Riesz F., Sz.-Nagy B., *Functional Analysis*, Blackie & Son Ltd., London–Glasgow, 1956.

[63] Slepian D., Pollak H.O., *Prolate Spheroidal Wave Functions, Fourier Analysis and Uncertainty, I*, Bell. Syst. Tech. J., 40, 43–63, 1961.

[64] Sakhnovich A.L., *On a Method of Inverting Toeplitz Matrices* (Russian), Mat. Issled., 8, 4, 180–186, 1973.

[65] Sakhnovich A.L., *Nonlinear Schrödinger Equation on a Semi-axis and an Inverse Problem Associated with it*, Ukr. Math. J., 42, 3, 316–323, 1990.

[66] Sakhnovich A.L., *The N-Wave Problem on the Semi-Axis*, Russ. Math. Surveys, 46, 4, 198–200, 1991.

[67] Sakhnovich A.L., *Skew-Self-Adjoint Discrete and Continuous Dirac-Type Systems: Inverse Problems and Borg–Marchenko Theorems*, Inverse Problems, 22, 2083–2101, 2006.

[68] Sakhnovich A.L., *Weyl functions, Inverse Problem and Special Solutions for the System Auxiliary to the Nonlinear Optics Equation*, Inverse Problems, 24, 025026, 2008.

[69] Sakhnovich A.L., Sakhnovich L.A., Roitberg I.Ya., *Inverse Problems and Nonlinear Evolution Equations. Solutions, Darboux Matrices and Weyl–Titchmarsh Functions*, de Gruyter Studies in Mathematics, vol. 47, De Gruyter, Berlin, 2013.

[70] Sakhnovich L.A., *Reduction of a Non-Selfadjoint Operator with Continuous Spectrum to Diagonal Form* (Russian), Uspekhi Mat. Nauk 13, 4(82), 193–196, 1958.

[71] Sakhnovich L.A., *The Reduction of Non-Selfadjoint Operators to Triangular Form* (Russian), Izv. Vyssh. Uchebn. Zaved. Mat., 1, 8, 180–186, 1959.

[72] Sakhnovich L.A., *A Study of the "Triangular Form" of Non-Selfadjoint Operators* (Russian), Izv. Vyssh. Uchebn. Zaved. Mat., 4, 11, 141–149, 1959. Translated in Amer. Math. Soc. Transl. (2), 54, 1966.

[73] Sakhnovich L.A., *Dissipative Operators with Absolutely Continuous Spectrum* (Russian), Tr. Mosk. Mat. O.-va, 19, 211–270, 1968. Translated in Trans. Moscow Math. Soc., 19, 223–297, 1968.

[74] Sakhnovich L.A., *Operators Similar to Unitary Operators with Absolutely Continuous Spectrum* (Russian), Funkcional. Anal. i Prilozhen., 2, 1, 51–63, 1968.

[75] Sakhnovich L.A., *The Similarity of Linear Operators* (Russian), Dokl. Akad. Nauk SSSR, 200, 541–544, 1971. Translated in Soviet Math. Dokl., 12, 5, 1445–1449, 1971.

[76] Sakhnovich L.A., *Similarity of Operators* (Russian), Sibirsk. Mat. Zh., 13, 868–883, 1972. Translated in Siberian Math. J., 13, 4 , 604–615, 1973.

[77] Sakhnovich L.A., *An Integral Equation with a Kernel Dependent on the Difference of the Arguments* (Russian), Mat. Issled. 8, 2, 138–146, 1973.

[78] Sakhnovich L.A., *On the Class of Integral Equations Solvable in the Exact Form* (Russian), Mat. Issled., 9, 1(31), 157–164, 1974.

[79] Sakhnovich L.A., *On the Factorization of an Operator-Valued Transfer Function* (Russian), Dokl. Akad. Nauk SSSR, 226, 4, 781–784, 1976. Translated in Soviet Math. Dokl. 17, 204–207, 1976.

[80] Sakhnovich L.A., *Factorization of Operators in $L^2(a,b)$*, in: *Investigations on linear operators and function theory. 99 unsolved problems in linear and complex analysis*, Research on linear operators and function theory, Zap. Nauchn. Sem. LOMI, 81, pp. 103–106, LOMI, Leningrad, 1978. Translated in J. Soviet Math., 26, 5, 2157–2159, 1984.

[81] Sakhnovich L.A., *Triangular Integro-Differential Operators with a Difference Kernel* (Russian), Sibirsk. Mat. Zh., 19, 4, 871–877, 1978.

[82] Sakhnovich L.A., *Factorization of Operators in $L^2(a,b)$* (Russian), Funkcional. Anal. i Prilozhen., 13, 3, 40–45, 1979. Translated in Funct. Anal. Appl., 13, 187–192, 1979.

[83] Sakhnovich L.A., *System of Equations with Difference Kernels* (Russian), Ukrain. Mat. Zh. 32, 1, 61–68, 1980. Translated in Ukrainian Mat. J. 32, 1, 44–50, 1980.

[84] Sakhnovich L.A., *Equations with a Difference Kernel on a Finite Interval* (Russian), Uspekhi Mat. Nauk, 35, 4, 69–129, 1980. Translated in Russ. Math. Survey, 35, 4, 81–152, 1980.

[85] Sakhnovich L.A., *The Asymptotic Behavior of the Spectrum of an Anharmonic Oscillator* (Russian), Teoret. Mat. Fiz., 47, 2, 266–276, 1981. Translated in Theoret. and Math. Phys., 47, 2, 449–456, 1981.

[86] Sakhnovich L.A., *On a Class of Integro-Differential Equations* (Russian), Ukrain. Mat. Zh., 34, 3, 328–333, 1982. Translated in Ukrainian Math. J. 34, 266–271, 1983.

[87] Sakhnovich L.A., *Abel Integral Equations in the Theory of Stable Processes* (Russian), Ukrain. Math. Zh., 36, 2, 213–218, 1983. Translated in Ukrainian Math. J., 36, 193–197, 1984.

[88] Sakhnovich L.A., *Factorization of Operators in $L^2(a,b)$*, in Havin V.P. et al. (eds.), *Linear and Complex Analysis Problem Book: 199 Research Problems*, Lecture Notes in Mathematics, vol. 1043, pp. 172–174, Springer, 1984.

[89] Sakhnovich L.A., *Equations with a Difference Kernel on a System of Segments* (Russian), Teor. Funkts., Funkts. Anal. Prilozh., 45, 111–122, 1986. Translated in J. Soviet Math., 48, 4, 464–475, 1990.

[90] Sakhnovich L.A., *Problems of Factorization and Operator Identities* (Russian), Uspekhi Mat. Nauk, 41, 1(247), 4–55, 1986. Translated in Russ. Math. Survey, 41, 1, 1–64, 1986.

[91] Sakhnovich L.A., *Nonlinear Equations and Inverse Problems on a Semi-Axis* (Russian), preprint, Akad. Nauk Ukrain. SSR, Inst. Mat., preprint no. 30, 55 p., 1987.

[92] Sakhnovich L.A., *Integral Equations with Difference Kernels on Finite Intervals*, 1st edition, Operator Theory Adv. Appl., vol. 84, Birkhäuser, Basel-Boston-Berlin, 1996.

[93] Sakhnovich L.A., *Interpolation Theory and its Applications*, Mathematics and its Applications, vol. 428, Kluwer Academic Publishers, Dordrecht, 1997.

[94] Sakhnovich L.A., *Spectral Theory of Canonical Differential Systems. Method of Operator Identities*, Operator Theory Adv. Appl., vol. 107, Birkhäuser, Basel, 1999.

[95] Sakhnovich L.A., *On Reducing the Canonical System to Two Dual Differential Systems*, J. Math. Anal. Appl., 255, 2, 499–509, 2001.

[96] Sakhnovich L.A., *On Triangular Factorization of Positive Operators*, in: Operator Theory Adv. Appl., vol. 179, pp. 289–308, Birkhäuser, Basel, 2007.

[97] Sakhnovich L.A., *Lévy Processes, Integral Equations, Statistical Physics: Connections and Interactions,* Operator Theory Adv. Appl., vol. 225, Birkhäuser, Basel, 2012.

[98] Sakhnovich L.A., *Lévy Processes: Long Time Behavior and Convolution-Type Form of the Itō Representation of the Infinitesimal Generator*, arXiv: 1306.1492

[99] Sato K., *Lévy Processes and Infinitely Divisible Distributions*, Cambridge Univ. Press, Cambridge, 1999.

[100] Sobolev V.V., *Scattering of Light in the Atmospheres of Planets* (Russian), Nauka, Moscow, 1972.

[101] Solodovnikov V.V., *Statistical Dynamics of Linear Automatic Control Systems* (Russian), Fizmatgiz, Moscow, 1960.

[102] Staerman I.Y., *Contact Problem of Theory of Elasticity* (Russian), Fizmatgiz, Moscow, 1949.

[103] Stone M., *Linear Transformation in Hilbert Space and Their Applications to Analysis,* Amer. Math. Soc. Coll. Publ., vol. 15, Amer. Math. Soc., Providence, 1932.

[104] Titchmarsh E.C., *Introduction to the Theory of Fourier Integrals*, Clarendon Press, Oxford, 1937.

[105] Tricomi F., *Integral Equations*, Interscience, New York-London, 1957.

[106] Thorin G.O., *Convexity Theorems Generalizing those of M. Riesz and Hadamard with some Applications*, Meddel. Lunds Univ. Mat. Semin., 9, 1948.

[107] Tucker H.G., *The Supports of Infinitely Divisible Distribution Functions*, Proc. Amer. Math. Soc., 49, 436–440, 1975.

[108] Van Casteren J.A., *Markov Processes, Feller Semigroups and Evolution Equations*, Series on concrete and applicable mathematics, vol. 12, World Scientific, Singapore, 2011.

[109] Vasil'eva I.A., Sakhnovich L.A., *Radiative Transfer in a Planar Gas Medium in the Presence of Scattering*, High Temperature, 25:3, 411–419, 1987.

[110] Vorovich I.I., Aleksandrov V.M., Babeshko V.A., *Non Classical Mixed Problems in Elasticity Theory* (Russian), Nauka, Moscow, 1974.

[111] Widom H., *Stable Processes and Integral Equations*, Trans. Amer. Math. Soc., 98, 430–449, 1961.

[112] Wiener N., *Extrapolation, Interpolation and Smoothing of Stationary Time Series*, Chapman & Hall Ltd, London, 1949.

[113] Zadeh L.A., Ragazzini J.R., *An Extension of Wiener's Theory of Prediction*, J. Appl. Physics, 21, 7, 645–655, 1950.

[114] Zygmund A., *Trigonometric Series, I, II*, Cambridge Univ. Press, Cambridge, 1959.

Glossary of Important Notations

$\vec{A}$	bounded triangular operator, see (11.1.1)
$A,\ A^*$	operators given by formulas (6) or (2.1.2)
$(A, \gamma, \nu(\,\mathrm{d}x))$	Lévy–Khinchine triplet, see (7.1.3)
$B(x, \lambda)$	solution of the equation $Sf = \mathrm{e}^{\mathrm{i}x\lambda}$, see (1.3.9)
$B_\gamma(x, \lambda)$	solution of the equation $Sf = \mathrm{e}^{\mathrm{i}x\lambda}$, see (2.1.6)
$\mathbb{C}$	complex plane
$\mathbb{C}_+$	open upper half-plane $\{z : \operatorname{Im}(z) > 0\}$
$\mathbb{C}_-$	open lower half-plane $\{z : \operatorname{Im}(z) < 0\}$
C_0	space of the continuous functions $f(x)$, which satisfy the condition $\lim f(x) = 0$ $(|x| \to \infty)$ and have the norm defined by $\|f\| = \sup_x |f(x)|$
C_0^n	set of the functions $f(x) \in C_0$ such that $f^{(k)}(x) \in C_0$ $(1 \le k \le n)$
$C(a)$	set of functions $f(x) \in C_0$ which satisfy (7.2.1)
C_Δ	special set of functions, see (8.3.16)
col	column, col $\begin{bmatrix} g_1 & g_2 \end{bmatrix} = \begin{bmatrix} g_1 \\ g_2 \end{bmatrix}$
$\mathfrak{D}$	set of generalized functions of the form (2.3.1)
$D(A, B)$	connected set of complex numbers, which contains $z = \infty$ but contains neither points of spectrum of A nor of B.
$D(T)$	domain of definition of the operator T
D_V	domain of definition of the operator V, see (9.2.2)
D_Δ	space of the continuous functions $g(x)$ on the domain Δ
D_Δ^0	subspace of D_Δ, see (8.4.1)
deg	order of a polynomial
diag	diagonal matrix
dim	dimension of a space
e_x	characteristic function, see (1.1.2)
e^z	exponential function, $\mathrm{e}^z = \exp(z)$
$E[\cdot]$	mathematical expectation
$F_0(x, t)$	distribution function, see (8.3.1)

gmul λ_k geometric multiplicity of the eigenvalue λ_k, i.e., a number
              of linearly independent eigenfunctions, corresponding
              to this eigenvalue

i   complex unity, $i^2 = -1$

I identity operator

I_n $n \times n$ identity matrix

Im   imaginary part of either complex number or matrix

ind λ_k index of the eigenvalue, i.e., the dimension of the largest
              Jordan block associated with this eigenvalue

$\mathcal{J}^{i\alpha}$ fractional integral operator of purely imaginary order $i\alpha$,
              see (1.0.4) and Section 11.4

$\mathcal{J}(\tau, \eta)$ total intensity of radiation, see (3.8.1)

$K_F(\mu_j)$ multiplicity of the root μ_j of $F(\mu)$, see Section 4.2 point 3

$K_G(\lambda_j)$ multiplicity of the root λ_j of $G(\lambda)$, see Section 4.2 point 3

ker A kernel of an operator A, that is, the subspace,
              which A maps to zero

$\mathcal{L}_m$ solution of the equation $S\mathcal{L}_m = x^{m-1}$, see (1.2.1)

$L(0, \omega)$ space $L^1(0, \omega)$

$L^p(0, \omega)$ normed space of functions f with the norm

$$\|f\|_p = \left(\int_0^\omega |f(x)|^p \, dx \right)^{1/p}$$

$L_n^p(0, \omega)$ normed space of vector functions f $(f(x) \in \mathbb{C}^n)$

              with the norm $\|f\|_p = \left(\int_0^\omega \left(f^*(x) f(x) \right)^{p/2} dx \right)^{1/p}$

L_Δ truncated generator, see (8.3.25)

M function, $M(x) = s(x)$ $(0 \le x \le \omega)$, see (7)

mes Δ the summarized length of the non-intersecting segments
              $[a_k, b_k]$ such that $a_1 < b_1 < a_2 < b_2 < \cdots < a_n < b_n$,
              i.e., mes $\Delta = \sum_{k=1}^n (b_k - a_k)$

N_1 solution of the equation $SN_1 = M$, see (10)

N_2 solution of the equation $SN_2 = \mathbb{1}$, see (10)

$p(t, \Delta)$ probability that a sample of X_τ remains
              inside Δ for $0 \le \tau \le t$, see (8.1.2)

$P(X = 0)$ probability of the event $X = 0$

P_t transition operator, see (7.1.5)

$P_t(x_0, \Delta)$ probability $P(X_t \in \Delta)$ when $P(X_0 = x_0) = 1$, see (7.1.5)

P_Δ projector, $P_\Delta f(x) = \begin{cases} f(x) & \text{for} \quad x \in \Delta, \\ 0 & \text{for} \quad x \notin \Delta. \end{cases}$

$\mathbb{R}$ real axis

R_V	range of the operator V, see (9.2.2)
range (A)	range of an operator A
Re	real part of either complex number or matrix
S^*	operator adjoint to S (or its extension), see, e.g., Proposition 2.1.1
sgn	function, $\mathrm{sgn}\, u = \begin{cases} 1, & u > 0, \\ 0, & u = 0, \\ -1, & u < 0 \end{cases}$
$w_A(z)$	transfer matrix function, see (5.1.4)
$W_p^{(l)}$	set of functions $\varphi(x)$ such that $\varphi^{(l)}(x) \in L^p(0,\omega)$, see Section 2.2
$\Gamma(z)$	gamma function
$\delta(x)$	delta function
κ	number of the negative eigenvalues of the operator $S - \nu I$ (counted with their multiplicities)
$\nu(\,\mathrm{d}x)$	Lévy measure satisfying (7.1.4)
$\rho(\lambda, \mu)$	function, see (12), (2.1.17)
φ^*	function $\varphi^* = U\varphi$, see (4.1.9)
$\Psi(x, s, u)$	Laplace transform of $F(x, t, u)$, see (8.3.10)
$\Psi_\infty(x, s)$	function, see (8.3.21)
$\{H_1, H_2\}$	class of linear bounded operators acting from the Banach space H_1 into the Banach space H_2
$\langle \cdot, \cdot \rangle$	scalar product, i.e., $\langle f, g \rangle = \int_0^\omega g(x)^* f(x)\,\mathrm{d}x$, see (1.1.3), (1.4.2), (5.2.32)
$\langle \cdot, \cdot \rangle_\Delta$	scalar product in $L^2(\Delta)$, that is, in L^2 on the set of segments Δ
$\langle \cdot, \cdot \rangle_H$	scalar product in the Hilbert space H
$\lfloor K \rfloor$	integer part of K
$\mathbb{1}$	function, which identically equals 1
$\mathbb{1}_{|x|<1}$	function of x, which equals 1 when $|x| < 1$ and equals 0 when $|x| > 1$
$\displaystyle\fint_0^\omega$	Cauchy integral, principal value

Index

Printed in the United States
By Bookmasters